# The Number System

# Landmarks in the Hundreds

## Grade 3

*Also appropriate for Grade 4*

Susan Jo Russell
Andee Rubin

*Developed at TERC, Cambridge, Massachusetts*

**Dale Seymour Publications®**
Menlo Park, California

The *Investigations* curriculum was developed at TERC (formerly
Technical Education Research Centers) in collaboration with Kent State
University and the State University of New York at Buffalo. The work was
supported in part by National Science Foundation Grant No. ESI-9050210.
TERC is a nonprofit company working to improve mathematics and science
education. TERC is located at 2067 Massachusetts Avenue, Cambridge,
MA 02140.

**This project was supported, in part,
by the**
**National Science Foundation**
Opinions expressed are those of the authors
and not necessarily those of the Foundation

Managing Editor: Catherine Anderson
Series Editor: Beverly Cory
Revision Team: Laura Marshall Alavosus, Ellen Harding, Patty Green Holubar,
Suzanne Knott, Beverly Hersh Lozoff
ESL Consultant: Nancy Sokol Green
Production/Manufacturing Director: Janet Yearian
Production/Manufacturing Coordinator: Amy Changar, Shannon Miller
Design Manager: Jeff Kelly
Design: Don Taka
Illustrations: Jane McCreary, Carl Yoshihara
Cover: Bay Graphics
Composition: Archetype Book Composition

This book is published by Dale Seymour Publications®, an imprint of
Addison Wesley Longman, Inc.

Dale Seymour Publications
2725 Sand Hill Road
Menlo Park, CA 94025
Customer Service: 800-872-1100

Order number DS43845
ISBN 1-57232-698-0
8 9 10 11-ML-05 04 03 02

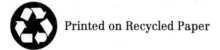

Printed on Recycled Paper

T E R C

*Principal Investigator*    Susan Jo Russell

*Co-Principal Investigator*    Cornelia C. Tierney

*Director of Research and Evaluation*    Jan Mokros

*Curriculum Development*

Joan Akers
Michael T. Battista
Mary Berle-Carman
Douglas H. Clements
Karen Economopoulos
Ricardo Nemirovsky
Andee Rubin
Susan Jo Russell
Cornelia C. Tierney
Amy Shulman Weinberg

*Evaluation and Assessment*

Mary Berle-Carman
Abouali Farmanfarmaian
Jan Mokros
Mark Ogonowski
Amy Shulman Weinberg
Tracey Wright
Lisa Yaffee

*Teacher Support*

Rebecca B. Corwin
Karen Economopoulos
Tracey Wright
Lisa Yaffee

*Technology Development*

Michael T. Battista
Douglas H. Clements
Julie Sarama Meredith
Andee Rubin

*Video Production*

David A. Smith

*Administration and Production*

Amy Catlin
Amy Taber

*Cooperating Classrooms
for This Unit*

Corrine Varon
Virginia M. Micciche
Joanne Walcott
Cambridge Public Schools
Cambridge, MA

Angela Philactos
Boston Public Schools
Boston, MA

Jeanne Wall
Linda Harding
Barbara Walsh
Arlington Public Schools
Arlington, MA

Katie Bloomfield
Robert A. Dihlmann
Shutesbury Elementary
Shutesbury, MA

*Consultants and Advisors*

Elizabeth Badger
Deborah Lowenberg Ball
Marilyn Burns
Ann Grady
Joanne M. Gurry
James J. Kaput
Steven Leinwand
Mary M. Lindquist
David S. Moore
John Olive
Leslie P. Steffe
Peter Sullivan
Grayson Wheatley
Virginia Woolley
Anne Zarinnia

*Graduate Assistants*

*Kent State University*
Joanne Caniglia
Pam DeLong
Carol King

*State University of New York at Buffalo*
Rosa Gonzalez
Sue McMillen
Julie Sarama Meredith
Sudha Swaminathan

*Revisions and Home Materials*

Cathy Miles Grant
Marlene Kliman
Margaret McGaffigan
Megan Murray
Kim O'Neil
Andee Rubin
Susan Jo Russell
Lisa Seyferth
Myriam Steinback
Judy Storeygard
Anna Suarez
Cornelia Tierney
Carol Walker
Tracey Wright

# CONTENTS

## TEACHER NOTES

## WHERE TO START

The first-time user of *Landmarks in the Hundreds* should read the following:

When you next teach this same unit, you can begin to read more of the background. Each time you present the unit, you will learn more about how your students understand the mathematical ideas.

*Investigations in Number, Data, and Space*® is a K–5 mathematics curriculum with four major goals:

- to offer students meaningful mathematical problems
- to emphasize depth in mathematical thinking rather than superficial exposure to a series of fragmented topics
- to communicate mathematics content and pedagogy to teachers
- to substantially expand the pool of mathematically literate students

The *Investigations* curriculum embodies a new approach based on years of research about how children learn mathematics. Each grade level consists of a set of separate units, each offering 2–8 weeks of work. These units of study are presented through investigations that involve students in the exploration of major mathematical ideas.

Approaching the mathematics content through investigations helps students develop flexibility and confidence in approaching problems, fluency in using mathematical skills and tools to solve problems, and proficiency in evaluating their solutions. Students also build a repertoire of ways to communicate about their mathematical thinking, while their enjoyment and appreciation of mathematics grows.

The investigations are carefully designed to invite all students into mathematics—girls and boys, members of diverse cultural, ethnic, and language groups, and students with different strengths and interests. Problem contexts often call on students to share experiences from their family, culture, or community. The curriculum eliminates barriers—such as work in isolation from peers, or emphasis on speed and memorization—that exclude some students from participating successfully in mathematics. The following aspects of the curriculum ensure that all students are included in significant mathematics learning:

- Students spend time exploring problems in depth.
- They find more than one solution to many of the problems they work on.

- They invent their own strategies and approaches, rather than relying on memorized procedures.
- They choose from a variety of concrete materials and appropriate technology, including calculators, as a natural part of their everyday mathematical work.
- They express their mathematical thinking through drawing, writing, and talking.
- They work in a variety of groupings—as a whole class, individually, in pairs, and in small groups.
- They move around the classroom as they explore the mathematics in their environment and talk with their peers.

While reading and other language activities are typically given a great deal of time and emphasis in elementary classrooms, mathematics often does not get the time it needs. If students are to experience mathematics in depth, they must have enough time to become engaged in real mathematical problems. We believe that a minimum of five hours of mathematics classroom time a week—about an hour a day—is critical at the elementary level. The plan and pacing of the *Investigations* curriculum is based on that belief.

We explain more about the pedagogy and principles that underlie these investigations in Teacher Notes throughout the units. For correlations of the curriculum to the NCTM Standards and further help in using this research-based program for teaching mathematics, see the following books:

- *Implementing the* Investigations in Number, Data, and Space® *Curriculum*
- *Beyond Arithmetic: Changing Mathematics in the Elementary Classroom* by Jan Mokros, Susan Jo Russell, and Karen Economopoulos

This book is one of the curriculum units for *Investigations in Number, Data, and Space*. In addition to providing part of a complete mathematics curriculum for your students, this unit offers information to support your own professional development. You, the teacher, are the person who will make this curriculum come alive in the classroom; the book for each unit is your main support system.

Although the curriculum does not include student textbooks, reproducible sheets for student work are provided in the unit and are also available as Student Activity Booklets. Students work actively with objects and experiences in their own environment and with a variety of manipulative materials and technology, rather than with a book of instruction and problems. We strongly recommend use of the overhead projector as a way to present problems, to focus group discussion, and to help students share ideas and strategies.

Ultimately, every teacher will use these investigations in ways that make sense for his or her particular style, the particular group of students, and the constraints and supports of a particular school environment. Each unit offers information and guidance for a wide variety of situations, drawn from our collaborations with many teachers and students over many years. Our goal in this book is to help you, a professional educator, implement this curriculum in a way that will give all your students access to mathematical power.

## Investigation Format

The opening two pages of each investigation help you get ready for the work that follows.

**What Happens** This gives a synopsis of each session or block of sessions.

**Mathematical Emphasis** This lists the most important ideas and processes students will encounter in this investigation.

**What to Plan Ahead of Time** These lists alert you to materials to gather, sheets to duplicate, transparencies to make, and anything else you need to do before starting.

---

### INVESTIGATION 2

# Using Landmarks to Solve Problems

#### What Happens

**Sessions 1, 2, and 3: Moving Beyond 100** After reviewing what they have found out about the factors of 100, students use this knowledge to explore larger numbers. Through a series of activities set up as choices, they explore the multiples of 100 with different tools, including money, the 300 chart, and the calculator. They solve problems involving multiplication and division by 20, 25, and other familiar factors, in the context of money. They examine the patterns in counting by 20's to 1000, and they begin to develop strategies for division using these patterns.

**Session 4: Solving Problems with Money** After a whole-class discussion of one multiplication problem, students work in pairs on multiplication and division problems that involve money. They work carefully on one, two, or three problems, writing and drawing to explain how they found their solutions.

**Sessions 5 and 6: Real-World Multiplying and Dividing** These sessions begin with an introduction to standard division notation. Students then apply their understanding of landmarks to multiplication and division problems in real-world contexts. Through a series of activities set up as choices, they solve given problems, find groups of objects in the classroom that are close to multiples of 100 (e.g., 30 students with 10 fingers each), and create and solve their own problems. They use multiplication and division notation to record their work, and learn to recognize a variety of symbols used to write down multiplication and division. They also become familiar with how to do multiplication and division on the calculator.

#### Mathematical Emphasis

- Using knowledge about factors of 100 to understand the structure of multiples of 100 (if there are four 25's in 100, there are twelve 25's in 300)

- Developing strategies to solve problems in multiplication and division situations by using knowledge of factors and multiples

- Estimating real quantities that are close to 200, 300, and 400

- Reading and using standard multiplication and division notation to record problems and answers

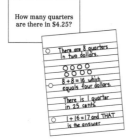

---

### INVESTIGATION 2

#### What to Plan Ahead of Time

##### Materials

- All materials used previously in the unit remain available: cubes, coins, and 100 charts. Make sure students know that they may use any of these materials that they would find helpful.

- Calculators: at least 1 per pair (all sessions)

- "Pictures of 100" that students made in Investigation 1 (Sessions 1–3)

- Overhead projector (Sessions 1–3)

##### Other Preparation

- Duplicate student sheets and teaching resources (located at the end of this unit) in the following quantities. If you have Student Activity Booklets, copy only the items marked with an asterisk, including any transparencies needed.

*For all sessions*
300 chart (p. 91): 5–7 per student (2 for homework), and 1 overhead transparency*

*For Sessions 1–3*
Student Sheet 11, Exploring Multiples of 100 (p. 79): 1–2 per student
Student Sheet 12, Calculator Skip Counting (p. 80): 1 per student

Money Problems* (p. 85): 5 copies, cut apart, and sets placed in boxes or envelopes
Student Sheet 13, How Many 4's Are in 600? (p. 81): 1 per student (homework)

*For Session 4*
More Money Problems* (p. 86): 5 copies, cut apart, and sets placed in boxes or envelopes

*For Sessions 5–6*
More Money Problems* (p. 86): 5 copies, cut apart, and sets placed in boxes or envelopes
Student Sheet 14, Multiplying Things in Class (p. 82): 1 per student, or 1 to post for student reference
Division Problems* (p. 87): 5 copies, cut apart, and sets placed in boxes or envelopes
Student Sheet 15, Multiplying Things at Home (p. 83): 1 per student (homework)
Student Sheet 16, Making Up Landmark Problems (p. 84): 1 per student (homework)

**Sessions** Within an investigation, the activities are organized by class session, a session being at least a one-hour math class. Sessions are numbered consecutively through an investigation. Often several sessions are grouped together, presenting a block of activities with a single major focus.

When you find a block of sessions presented together—for example, Sessions 1, 2, and 3—read through the entire block first to understand the overall flow and sequence of the activities. Make some preliminary decisions about how you will divide the activities into three sessions for your class, based on what you know about your students. You may need to modify your initial plans as you progress through the activities, and you may want to make notes in the margins of the pages as reminders for the next time you use the unit.

Be sure to read the Session Follow-Up section at the end of the session block to see what homework assignments and extensions are suggested as you make your initial plans.

While you may be used to a curriculum that tells you exactly what each class session should cover, we have found that the teacher is in a better position to make these decisions. Each unit is flexible and may be handled somewhat differently by every teacher. While we provide guidance for how many sessions a particular group of activities is likely to need, we want you to be active in determining an appropriate pace and the best transition points for your class. It is not unusual for a teacher to spend more or less time than is proposed for the activities.

**Ten-Minute Math** At the beginning of some sessions, you will find Ten-Minute Math activities. These are designed to be used in tandem with the investigations, but not during the math hour. Rather, we hope you will do them whenever you have a spare 10 minutes—maybe before lunch or recess, or at the end of the day.

Ten-Minute Math offers practice in key concepts, but not always those being covered in the unit. For example, in a unit on using data, Ten-Minute Math might revisit geometric activities done earlier in the year. Complete directions for the suggested activities are included at the end of each unit.

---

**Sessions 1, 2, and 3**

# Moving Beyond 100

**Materials**

- Students' pictures of 100
- Cubes, coins, and 100 charts
- Money Problems, in sets
- Student Sheet 11 (1–2 per student)
- Student Sheet 12 (1 per student)
- Transparency of 300 chart
- Overhead projector
- 300 chart (4–5 per student, 2 for homework)
- Calculators
- Student Sheet 13 (1 per student, homework)

**What Happens**

After reviewing what they have found out about the factors of 100, students use this knowledge to explore larger numbers. Through a series of activities set up as choices, they explore the multiples of 100 with different tools, including money, the 300 chart, and the calculator. They solve problems involving multiplication and division by 20, 25, and other familiar factors, in the context of money. They examine the patterns in counting by 20's to 1000, and they begin to develop strategies for division using these patterns. Their work focuses on:

- using what they know about 100 to think about multiples of 100
- proving how many groupings of a particular factor make 100

**Ten-Minute Math: Calendar Math** Once or twice during the next few days, do Calendar Math. Remember, Ten-Minute Math activities are designed to be done outside of the math hour.

Ask students to give number combinations that equal the number of the day's date, using multiplication as part of their answer.

List student responses. For example, if the date is January 21, solutions could include $2 \times 10 + 1$, $3 \times 10 - 9$, or $5 \times 5 - 4$.

Choose a "favorite expression" for the day, perhaps the most unusual, or one that uses a new idea. Use this favorite to write the date on the board, for example: January $0 \times 21 + 21$.

For variations, see p. 63.

**Activity**

**Hundreds from Home**

We've been working with the number 100 recently, and we're going to be working with even higher numbers soon. Let's start today by looking at the pictures of 100 you made earlier.

Ask several students to show the arrangements of 100 objects or pictures they made at school and at home. Have them describe how they arranged the items to be able to "prove" that there are exactly 100 objects without counting them one by one.

Spend a bit of time adding by 100's, using the student displays as concrete objects to add. See the **Dialogue Box**, Hundreds from Home (p. 39), for an example. Ask questions such as:

How many 20's are there altogether if we put Jamal's and Annie's together?

32 ■ *Investigation 2: Using Landmarks to Solve Problems*

---

**Activities** The activities include pair and small-group work, individual tasks, and whole-class discussions. In any case, students are seated together, talking and sharing ideas during all work times. Students most often work cooperatively, although each student may record work individually.

**Choice Time** In some units, some sessions are structured with activity choices. In these cases, students may work simultaneously on different activities focused on the same mathematical ideas. Students choose which activities they want to do, and they cycle through them.

You will need to decide how to set up and introduce these activities and how to let students make their choices. Some teachers present them as station activities, in different parts of the room. Some list the choices on the board as reminders or have students keep their own lists.

**Extensions** Sometimes in Session Follow-Up, you will find suggested extension activities. These are opportunities for some or all students to explore a topic in greater depth or in a different context. They are not designed for "fast" students; mathematics is a multifaceted discipline, and different students will want to go further in different investigations. Look for and encourage the sparks of interest and enthusiasm you see in your students, and use the extensions to help them pursue these interests.

**Excursions** Some of the *Investigations* units include excursions—blocks of activities that could be omitted without harming the integrity of the unit. This is one way of dealing with the great depth and variety of elementary mathematics— much more than a class has time to explore in any one year. Excursions give you the flexibility to make different choices from year to year, doing the excursion in one unit this time, and next year trying another excursion.

**Tips for the Linguistically Diverse Classroom** At strategic points in each unit, you will find concrete suggestions for simple modifications of the teaching strategies to encourage the participation of all students. Many of these tips offer alternative ways to elicit critical thinking from students at varying levels of English proficiency, as well as from other students who find it difficult to verbalize their thinking.

The tips are supported by suggestions for specific vocabulary work to help ensure that all students can participate fully in the investigations. The Preview for the Linguistically Diverse Classroom (p. I-19) lists important words that are assumed as part of the working vocabulary of the unit. Second-language learners will need to become familiar with these words in order to understand the problems and activities they will be doing. These terms can be incorporated into students' second-language work before or during the unit. Activities that can be used to present the words are found in the appendix, Vocabulary Support for Second-Language Learners (p. 65). In addition, ideas for making connections to students' language and cultures, included on the Preview page, help the class explore the unit's concepts from a multicultural perspective.

## Materials

A complete list of the materials needed for teaching this unit is found on p. I-16. Some of these materials are available in kits for the *Investigations* curriculum. Individual items can also be purchased from school supply dealers.

**Classroom Materials** In an active mathematics classroom, certain basic materials should be available at all times: interlocking cubes, pencils, unlined paper, graph paper, calculators, things to count with, and measuring tools. Some activities in this curriculum require scissors and glue sticks or tape. Stick-on notes and large paper are also useful materials throughout.

So that students can independently get what they need at any time, they should know where these materials are kept, how they are stored, and how they are to be returned to the storage area. For example, interlocking cubes are best stored in towers of ten; then, whatever the activity, they should be returned to storage in groups of ten at the end of the hour. You'll find that establishing such routines at the beginning of the year is well worth the time and effort.

**Technology** Calculators are used throughout *Investigations*. Many of the units recommend that you have at least one calculator for each pair. You will find calculator activities, plus Teacher Notes discussing this important mathematical tool, in an early unit at each grade level. It is assumed that calculators will be readily available for student use.

Computer activities at grade 3 use two software programs that were developed especially for the *Investigations* curriculum. *Tumbling Tetrominoes* is introduced in the 2-D Geometry unit, *Flips, Turns, and Area*. This game emphasizes ideas about area and about geometric motions (slides, flips, and turns). The program *Geo-Logo*™ is introduced in a second 2-D Geometry unit, *Turtle Paths*, where students use it to explore geometric shapes.

How you use the computer activities depends on the number of computers you have available. Suggestions are offered in the geometry units for how to organize different types of computer environments.

**Children's Literature** Each unit offers a list of suggested children's literature (p. I-16) that can be used to support the mathematical ideas in the unit. Sometimes an activity is based on a specific children's book, with suggestions for substitutions where practical. While such activities can be adapted and taught without the book, the literature offers a rich introduction and should be used whenever possible.

**Student Sheets and Teaching Resources** Student recording sheets and other teaching tools needed for both class and homework are provided as reproducible blackline masters at the end of each unit. They are also available as Student Activity Booklets. These booklets contain all the sheets each student will need for individual work, freeing you from extensive copying (although you may need or want to copy the occasional teaching resource on transparency film or card stock, or make extra copies of a student sheet).

We think it's important that students find their own ways of organizing and recording their work. They need to learn how to explain their thinking with both drawings and written words, and how to organize their results so someone else can

understand them. For this reason, we deliberately do not provide student sheets for every activity. Regardless of the form in which students do their work, we recommend that they keep a mathematics notebook or folder so that their work is always available for reference.

**Homework** In *Investigations,* homework is an extension of classroom work. Sometimes it offers review and practice of work done in class, sometimes preparation for upcoming activities, and sometimes numerical practice that revisits work in earlier units. Homework plays a role both in supporting students' learning and in helping inform families about the ways in which students in this curriculum work with mathematical ideas.

Depending on your school's homework policies and your own judgment, you may want to assign more homework than is suggested in the units. For this purpose you might use the practice pages, included as blackline masters at the end of this unit, to give students additional work with numbers.

---

Name _____ Date _____

**Ways to Split Up a Dollar**

| Number of People Sharing a Dollar | How Much Each Person Gets |
|---|---|
|  |  |
|  |  |
|  |  |
|  |  |
|  |  |
|  |  |
|  |  |
|  |  |
|  |  |
|  |  |
|  |  |
|  |  |
| Cannot split a dollar evenly: | |

© Dale Seymour Publications®          77          *Investigation 1 • Sessions 6–7 Landmarks in the Hundreds*

For some homework assignments, you will want to adapt the activity to meet the needs of a variety of students in your class: those with special needs, those ready for more challenge, and second-language learners. You might change the numbers in a problem, make the activity more or less complex, or go through a sample activity with those who need extra help. You can modify any student sheet for either homework or class use. In particular, making numbers in a problem smaller or larger can make the same basic activity appropriate for a wider range of students.

Another issue to consider is how to handle the homework that students bring back to class—how to recognize the work they have done at home without spending too much time on it. Some teachers hold a short group discussion of different approaches to the assignment; others ask students to share and discuss their work with a neighbor, or post the homework around the room and give students time to tour it briefly. If you want to keep track of homework students bring in, be sure it ends up in a designated place.

*Investigations* at Home  It is a good idea to make your policy on homework explicit to both students and their families when you begin teaching with *Investigations*. How frequently will you be assigning homework? When do you expect homework to be completed and brought back to school? What are your goals in assigning homework? How independent should families expect their children to be? What should the parent's or guardian's role be? The more explicit you can be about your expectations, the better the homework experience will be for everyone.

*Investigations* at Home (a booklet available separately for each unit, to send home with students) gives you a way to communicate with families about the work students are doing in class. This booklet includes a brief description of every session, a list of the mathematics content emphasized in each investigation, and a discussion of each homework assignment to help families more effectively support their children. Whether or not you are using the *Investigations* at Home booklets, we expect you to make your own choices about home-

work assignments. Feel free to omit any and to add extra ones you think are appropriate.

**Family Letter**  A letter that you can send home to students' families is included with the blackline masters for each unit. Families need to be informed about the mathematics work in your classroom; they should be encouraged to participate in and support their children's work. A reminder to send home the letter for each unit appears in one of the early investigations. These letters are also available separately in Spanish, Vietnamese, Cantonese, Hmong, and Cambodian.

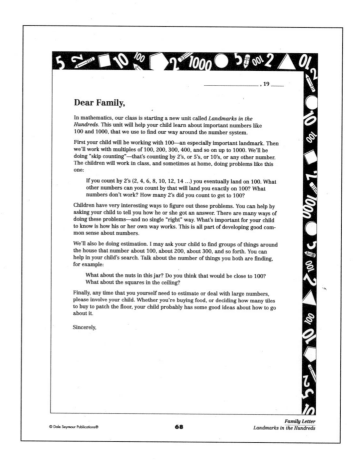

## Help for You, the Teacher

Because we believe strongly that a new curriculum must help teachers think in new ways about mathematics and about their students' mathematical thinking processes, we have included a great deal of material to help you learn more about both.

**About the Mathematics in This Unit**  This introductory section (p. I-17) summarizes the critical information about the mathematics you will be teaching. It describes the unit's central mathematical ideas and how students will encounter them through the unit's activities.

**Teacher Notes**  These reference notes provide practical information about the mathematics you are teaching and about our experience with how students learn. Many of the notes were written in response to actual questions from teachers, or to discuss important things we saw happening in the field-test classrooms. Some teachers like to read them all before starting the unit, then review them as they come up in particular investigations.

**Dialogue Boxes**  Sample dialogues demonstrate how students typically express their mathematical ideas, what issues and confusions arise in their thinking, and how some teachers have guided class discussions.

These dialogues are based on the extensive classroom testing of this curriculum; many are word-for-word transcriptions of recorded class discussions. They are not always easy reading; sometimes it may take some effort to unravel what the students are trying to say. But this is the value of these dialogues; they offer good clues to how your students may develop and express their approaches and strategies, helping you prepare for your own class discussions.

**Where to Start**  You may not have time to read everything the first time you use this unit. As a first-time user, you will likely focus on understanding the activities and working them out with your students. Read completely through each investigation before starting to present it. Also read those sections listed in the Contents under the heading Where to Start (p. vi).

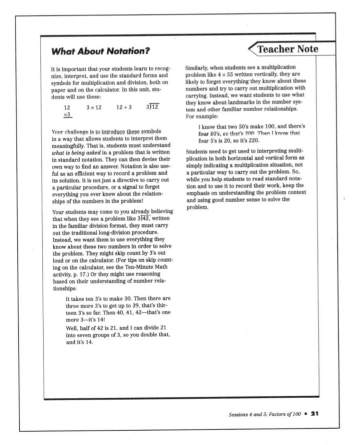

The *Investigations* curriculum incorporates the use of two forms of technology in the classroom: calculators and computers. Calculators are assumed to be standard classroom materials, available for student use in any unit. Computers are explicitly linked to one or more units at each grade level; they are used with the unit on 2-D geometry at each grade, as well as with some of the units on measuring, data, and changes.

## Using Calculators

In this curriculum, calculators are considered tools for doing mathematics, similar to pattern blocks or interlocking cubes. Just as with other tools, students must learn both *how* to use calculators correctly and *when* they are appropriate to use. This knowledge is crucial for daily life, as calculators are now a standard way of handling numerical operations, both at work and at home.

Using a calculator correctly is not a simple task; it depends on a good knowledge of the four operations and of the number system, so that students can select suitable calculations and also determine what a reasonable result would be. These skills are the basis of any work with numbers, whether or not a calculator is involved.

Unfortunately, calculators are often seen as tools to check computations with, as if other methods are somehow more fallible. Students need to understand that any computational method can be used to check any other; it's just as easy to make a mistake on the calculator as it is to make a mistake on paper or with mental arithmetic. Throughout this curriculum, we encourage students to solve computation problems in more than one way in order to double-check their accuracy. We present mental arithmetic, paper-and-pencil computation, and calculators as three possible approaches.

In this curriculum we also recognize that, despite their importance, calculators are not always appropriate in mathematics instruction. Like any tools, calculators are useful for some tasks, but not for others. You will need to make decisions about when to allow students access to calculators and when to ask that they solve problems without them, so that they can concentrate on other tools and skills. At times when calculators are or are not appropriate for a particular activity, we make specific recommendations. Help your students develop their own sense of which problems they can tackle with their own reasoning and which ones might be better solved with a combination of their own reasoning and the calculator.

Managing calculators in your classroom so that they are a tool, and not a distraction, requires some planning. When calculators are first introduced, students often want to use them for everything, even problems that can be solved quite simply by other methods. However, once the novelty wears off, students are just as interested in developing their own strategies, especially when these strategies are emphasized and valued in the classroom. Over time, students will come to recognize the ease and value of solving problems mentally, with paper and pencil, or with manipulatives, while also understanding the power of the calculator to facilitate work with larger numbers.

Experience shows that if calculators are available only occasionally, students become excited and distracted when they are permitted to use them. They focus on the tool rather than on the mathematics. In order to learn when calculators are appropriate and when they are not, students must have easy access to them and use them routinely in their work.

If you have a calculator for each student, and if you think your students can accept the responsibility, you might allow them to keep their calculators with the rest of their individual materials, at least for the first few weeks of school. Alternatively, you might store them in boxes on a shelf, number each calculator, and assign a corresponding number to each student. This system can give students a sense of ownership while also helping you keep track of the calculators.

## Using Computers

Students can use computers to approach and visualize mathematical situations in new ways. The computer allows students to construct and manipulate geometric shapes, see objects move according to rules they specify, and turn, flip, and repeat a pattern.

This curriculum calls for computers in units where they are a particularly effective tool for learning mathematics content. One unit on 2-D geometry at each of the grades 3–5 includes a core of activities that rely on access to computers, either in the classroom or in a lab. Other units on geometry, measurement, data, and changes include computer activities, but can be taught without them. In these units, however, students' experience is greatly enhanced by computer use.

The following list outlines the recommended use of computers in this curriculum:

### Grade 1
Unit: *Survey Questions and Secret Rules*
   (Collecting and Sorting Data)
Software: Tabletop, Jr.
Source: Broderbund

Unit: *Quilt Squares and Block Towns*
   (2-D and 3-D Geometry)
Software: *Shapes*
Source: provided with the unit

### Grade 2
Unit: *Mathematical Thinking at Grade 2*
   (Introduction)
Software: *Shapes*
Source: provided with the unit

Unit: *Shapes, Halves, and Symmetry*
   (Geometry and Fractions)
Software: *Shapes*
Source: provided with the unit

Unit: *How Long? How Far?* (Measuring)
Software: *Geo-Logo*
Source: provided with the unit

### Grade 3
Unit: *Flips, Turns, and Area* (2-D Geometry)
Software: *Tumbling Tetrominoes*
Source: provided with the unit

Unit: *Turtle Paths* (2-D Geometry)
Software: *Geo-Logo*
Source: provided with the unit

### Grade 4
Unit: *Sunken Ships and Grid Patterns*
   (2-D Geometry)
Software: *Geo-Logo*
Source: provided with the unit

### Grade 5
Unit: *Picturing Polygons* (2-D Geometry)
Software: *Geo-Logo*
Source: provided with the unit

Unit: *Patterns of Change* (Tables and Graphs)
Software: *Trips*
Source: provided with the unit

Unit: *Data: Kids, Cats, and Ads* (Statistics)
Software: Tabletop, Sr.
Source: Broderbund

The software provided with the *Investigations* units uses the power of the computer to help students explore mathematical ideas and relationships that cannot be explored in the same way with physical materials. With the *Shapes* (grades 1–2) and *Tumbling Tetrominoes* (grade 3) software, students explore symmetry, pattern, rotation and reflection, area, and characteristics of 2-D shapes. With the *Geo-Logo* software (grades 3–5), students investigate rotations and reflections, coordinate geometry, the properties of 2-D shapes, and angles. The *Trips* software (grade 5) is a mathematical exploration of motion in which students run experiments and interpret data presented in graphs and tables.

We suggest that students work in pairs on the computer; this not only maximizes computer resources but also encourages students to consult, monitor, and teach one another. Generally, more than two students at one computer find it difficult to share. Managing access to computers is an issue for every classroom. The curriculum gives you explicit support for setting up a system. The units are structured on the assumption that you have enough computers for half your students to work on the machines in pairs at one time. If you do not have access to that many computers, suggestions are made for structuring class time to use the unit with five to eight computers, or even with fewer than five.

Assessment plays a critical role in teaching and learning, and it is an integral part of the *Investigations* curriculum. For a teacher using these units, assessment is an ongoing process. You observe students' discussions and explanations of their strategies on a daily basis and examine their work as it evolves. While students are busy recording and representing their work, working on projects, sharing with partners, and playing mathematical games, you have many opportunities to observe their mathematical thinking. What you learn through observation guides your decisions about how to proceed. In any of the units, you will repeatedly consider questions like these:

- Do students come up with their own strategies for solving problems, or do they expect others to tell them what to do? What do their strategies reveal about their mathematical understanding?

- Do students understand that there are different strategies for solving problems? Do they articulate their strategies and try to understand other students' strategies?

- How effectively do students use materials as tools to help with their mathematical work?

- Do students have effective ideas for keeping track of and recording their work? Does keeping track of and recording their work seem difficult for them?

You will need to develop a comfortable and efficient system for recording and keeping track of your observations. Some teachers keep a clipboard handy and jot notes on a class list or on adhesive labels that are later transferred to student files. Others keep loose-leaf notebooks with a page for each student and make weekly notes about what they have observed in class.

## Assessment Tools in the Unit

With the activities in each unit, you will find questions to guide your thinking while observing the students at work. You will also find two built-in assessment tools: Teacher Checkpoints and embedded Assessment activities.

**Teacher Checkpoints** The designated Teacher Checkpoints in each unit offer a time to "check in" with individual students, watch them at work, and ask questions that illuminate how they are thinking.

At first it may be hard to know what to look for, hard to know what kinds of questions to ask. Students may be reluctant to talk; they may not be accustomed to having the teacher ask them about their work, or they may not know how to explain their thinking. Two important ingredients of this process are asking students open-ended questions about their work and showing genuine interest in how they are approaching the task. When students see that you are interested in their thinking and are counting on them to come up with their own ways of solving problems, they may surprise you with the depth of their understanding.

Teacher Checkpoints also give you the chance to pause in the teaching sequence and reflect on how your class is doing overall. Think about whether you need to adjust your pacing: Are most students fluent with strategies for solving a particular kind of problem? Are they just starting to formulate good strategies? Or are they still struggling with how to start? Depending on what you see as the students work, you may want to spend more time on similar problems, change some of the problems to use smaller numbers, move quickly to more challenging material, modify subsequent activities for some students, work on particular ideas with a small group, or pair students who have good strategies with those who are having more difficulty.

**Embedded Assessment Activities** Assessment activities embedded in each unit will help you examine specific pieces of student work, figure out what it means, and provide feedback. From the students' point of view, these assessment activities are no different from any others. Each is a learning experience in and of itself, as well as an opportunity for you to gather evidence about students' mathematical understanding.

The embedded assessment activities sometimes involve writing and reflecting; at other times, a discussion or brief interaction between student and teacher; and in still other instances, the creation and explanation of a product. In most cases, the assessments require that students *show* what they did, *write* or *talk* about it, or do both. Having to explain how they worked through a problem helps students be more focused and clear in their mathematical thinking. It also helps them realize that doing mathematics is a process that may involve tentative starts, revising one's approach, taking different paths, and working through ideas.

Teachers often find the hardest part of assessment to be interpreting their students' work. We provide guidelines to help with that interpretation. If you have used a process approach to teaching writing, the assessment in *Investigations* will seem familiar. For many of the assessment activities, a Teacher Note provides examples of student work and a commentary on what it indicates about student thinking.

### Documentation of Student Growth

To form an overall picture of mathematical progress, it is important to document each student's work in journals, notebooks, or portfolios. The choice is largely a matter of personal preference; some teachers have students keep a notebook or folder for each unit, while others prefer one mathematics notebook, or a portfolio of selected work for the entire year. The final activity in each *Investigations* unit, called Choosing Student Work to Save, helps you and the students select representative samples for a record of their work.

This kind of regular documentation helps you synthesize information about each student as a mathematical learner. From different pieces of evidence, you can put together the big picture. This synthesis will be invaluable in thinking about where to go next with a particular child, deciding where more work is needed, or explaining to parents (or other teachers) how a child is doing.

If you use portfolios, you need to collect a good balance of work, yet avoid being swamped with an overwhelming amount of paper. Following are some tips for effective portfolios:

■ Collect a representative sample of work, including some pieces that students themselves select for inclusion in the portfolio. There should be just a few pieces for each unit, showing different kinds of work—some assignments that involve writing, as well as some that do not.

■ If students do not date their work, do so yourself so that you can reconstruct the order in which pieces were done.

■ Include your reflections on the work. When you are looking back over the whole year, such comments are reminders of what seemed especially interesting about a particular piece; they can also be helpful to other teachers and to parents. Older students should be encouraged to write their own reflections about their work.

### Assessment Overview

There are two places to turn for a preview of the assessment opportunities in each *Investigations* unit. The Assessment Resources column in the unit Overview Chart (pp. I-13–I-15) identifies the Teacher Checkpoints and Assessment activities embedded in each investigation, guidelines for observing the students that appear within classroom activities, and any Teacher Notes and Dialogue Boxes that explain what to look for and what types of student responses you might expect to see in your classroom. Additionally, the section About the Assessment in This Unit (p. I-18) gives you a detailed list of questions for each investigation, keyed to the mathematical emphases, to help you observe student growth.

Depending on your situation, you may want to provide additional assessment opportunities. Most of the investigations lend themselves to more frequent assessment, simply by having students do more writing and recording while they are working.

# Landmarks in the Hundreds

**Content of This Unit**  Students work with 100, factors of 100, and multiples of 100 (up to 1000). They use coins, 100 charts, and cubes to investigate these numbers. They develop a sense of these quantities by counting real objects and by making "pictures" of 100 and 1000. They develop their own strategies, based on what they know about these landmark numbers, to solve multiplication and division problems.

**Connections with Other Units**  If you are doing the full-year *Investigations* curriculum in the suggested sequence for grade 3, this is the fifth of ten units. Your class will have already completed the Multiplication and Division unit *Things That Come in Groups.* There, students practiced skip counting by 2's, 3's, 4's, 5's, and 6's, and worked on such problems as "how many 4's in 48?" If your students have not had comparable experience, spend more time on the first three sessions of Investigation 1. During Investigation 2, these students may also need to spend more time working with multiples of 10 and with the smaller multiples of 100.

If your school is not using the full-year curriculum, this unit can also be used successfully at grade 4. The work in this unit is continued and extended in the grade 4–5 unit, *Landmarks in the Thousands.*

## *Investigations* Curriculum ■ Suggested Grade 3 Sequence

*Mathematical Thinking at Grade 3* (Introduction)

*Things That Come in Groups* (Multiplication and Division)

*Flips, Turns, and Area* (2-D Geometry)

*From Paces to Feet* (Measuring and Data)

▶ *Landmarks in the Hundreds* (The Number System)

*Up and Down the Number Line* (Changes)

*Combining and Comparing* (Addition and Subtraction)

*Turtle Paths* (2-D Geometry)

*Fair Shares* (Fractions)

*Exploring Solids and Boxes* (3-D Geometry)

# Investigation 1 ■ Finding Factors

| Class Sessions | Activities | Pacing |
|---|---|---|
| Session 1 (p. 4)<br>SKIP COUNTING WITH CUBES | Ways to Count to 10<br>Finding Factors with Cubes<br>Homework: Ways to Make _____ | minimum<br>1 hr |
| Sessions 2 and 3 (p. 7)<br>FACTORS OF 24, 36, AND 48 | Skip Counting on the 100 Chart<br>What's a Factor?<br>Making Counting Charts for 24<br>Factors of 36 and 48<br>Teacher Checkpoint: Showing Factors of 36 or 48<br>Class Discussion: What Did You Find Out About<br>   24, 36, and 48?<br>Homework: Factors of 36/Factors of 48 | minimum<br>2 hr |
| Sessions 4 and 5 (p. 17)<br>FACTORS OF 100 | What Numbers Are Factors of 100?<br>Making a Picture of 100<br>Homework: A Picture of 100 | minimum<br>2 hr |
| Sessions 6 and 7 (p. 22)<br>DIVIDING A DOLLAR | Listing the Factors of 100<br>Working with Coins<br>Dividing a Dollar Among Five People<br>How Can We Divide a Dollar Evenly?<br>Teacher Checkpoint: Factors of 100<br>Homework: Share a Dollar | minimum<br>2 hr |

◗ Ten-Minute Math ■ Counting Around the Class

## Mathematical Emphasis

- Understanding the relationship between skip counting and grouping (for example, as we count 3, 6, 9, 12, we are adding a group of 3 to the total each time)

- Becoming familiar with the relationships among commonly encountered factors and multiples (for example, is 3 a factor of 24? How many 3's does it take to make 24?)

- Increasing fluency in counting by single-digit numbers (2's, 3's, 4's, 6's, 8's) as well as by useful two-digit numbers (10's, 20's, 25's)

- Developing familiarity with the factors of 100, an important landmark in our number system, and their relationships to 100 through work with cubes, coins, and 100 charts

## Assessment Resources

Teacher Checkpoint: Showing Factors of 36 or 48 (p. 12)

Students' Problems with Skip Counting (Teacher Note, p. 14)

Teacher Checkpoint: Factors of 100 (p. 27)

## Materials

Interlocking cubes

Overhead projector and transparencies

Colored paper

Scissors

Glue or paste

Crayons or markers

Scrap materials

Plastic or real coins

Student Sheets 1–10

Teaching resource sheets

Family letter

*Two Ways to Count to Ten,* retold by Ruby Dee (opt.)

# Investigation 2 ▪ Using Landmarks to Solve Problems

| Class Sessions | Activities | Pacing |
|---|---|---|
| Sessions 1, 2, and 3 (p. 32)<br>MOVING BEYOND 100 | Hundreds from Home<br>Choice Time: Exploring Multiples of 100<br>Class Discussion: Finding Patterns in the 20's<br>Assessment: How Many in 500?<br>Homework: Money Problems<br>Homework: Skip Counting<br>Homework: How Many 4's Are in 600? | minimum<br>3 hr |
| Session 4 (p. 41)<br>SOLVING PROBLEMS WITH MONEY | Class Discussion: Six People Paying $1.50 Each<br>Problems Using Money<br>Homework: More Money Problems | minimum<br>1 hr |
| Sessions 5 and 6 (p. 44)<br>REAL-WORLD MULTIPLYING<br>AND DIVIDING | Using Standard Notation for Division<br>Choice Time: Working with Landmarks<br>Multiplying and Dividing on the Calculator<br>Homework: Multiplying Things at Home<br>Homework: Making Up Landmark Problems<br>Extension: Things That Number in the Low Hundreds<br>Extension: Things That Number in the High Hundreds<br>Extension: Exploring Different Calculators | minimum<br>2 hr |

◔ **Ten-Minute Math** ▪ **Calendar Math**

## Mathematical Emphasis

- Using knowledge about factors of 100 to understand the structure of multiples of 100 (if there are four 25's in 100, there are twelve 25's in 300)

- Developing strategies to solve problems in multiplication and division situations by using knowledge of factors and multiples

- Estimating real quantities that are close to 200, 300, and 400

- Reading and using standard multiplication and division notation to record problems and answers

## Assessment Resources

Assessment: How Many in 500? (Teacher Note, p. 37)

Hundreds from Home (Dialogue Box, p. 39)

How Many 20's in 280? (Dialogue Box, p. 40)

Using Multiples to Count (Teacher Note, p. 49)

Talking and Writing About Division (Teacher Note, p. 50)

## Materials

Interlocking cubes

Plastic or real coins

Calculators

Students' "Pictures of 100"

Overhead projector

Student Sheets 11–16

Teaching resource sheets

## Investigation 3 ▪ Constructing a 1000 Chart

| Class Sessions | Activities | Pacing |
|---|---|---|
| Session 1 (p. 54)<br>A 1000 CHART | Making a Thousand<br>Homework: More Than 300 | minimum<br>1 hr |
| Sessions 2 and 3 (p. 56)<br>FINDING LARGE QUANTITIES | Locating Numbers on the 1000 Chart<br>Finding Real Quantities for the 1000 Chart<br>Assessment: Calculating with the 1000 Chart<br>Choosing Student Work to Save | minimum<br>2 hr |

◗ **Ten-Minute Math ▪ Counting Around the Class**

### Mathematical Emphasis

▪ Using factors of 100 to understand the structure of 1000 (How many 50's does it take to make 1000?)

▪ Estimating quantities up to 1000 (What can we find in the classroom that numbers about 500?)

▪ Using landmarks to calculate "distances" within 1000 (How far is it from 650 to 950?)

### Assessment Resources

Assessment: Calculating with the 1000 Chart (p. 58)

Assessment: How Far from 650 to 1000? (Teacher Note, p. 59)

Choosing Student Work to Save (p. 59)

### Materials

Large poster or chart paper
Scissors
Glue or tape
Crayons or markers
Small counters or cubes
String or ribbon (opt.)
Interlocking cubes
Plastic or real coins
Calculators
Teaching resource sheets
Student Sheet 17

Following are the basic materials needed for the activities in this unit. Many of the items can be purchased from the publisher, either individually or in the Teacher Resource Package and the Student Materials Kit for grade 3. Detailed information is available on the *Investigations* order form. To obtain this form, call toll-free 1-800-872-1100 and ask for a Dale Seymour customer service representative.

Snap™ Cubes (interlocking cubes): at least 50 per pair of students

Plastic coins: at least 50 pennies, 25 nickels, 12 dimes, and 6 quarters per small group of students (real coins may be substituted or used in addition to these)

Calculators: at least 1 per pair of students

Small counters or cubes (1 cm): 1 per student

*Two Ways to Count to Ten*, retold by Ruby Dee (optional)

Large poster or chart paper: 1 sheet per pair of students

Colored paper

Scrap materials

Scissors

Glue or paste

Tape

Crayons or markers

String or ribbon (optional)

Overhead projector

The following materials are provided at the end of this unit as blackline masters. A Student Activity Booklet containing all student sheets and teaching resources needed for individual work is available.

Family Letter (p. 68)

Student Sheets 1–17 (p. 69)

Teaching Resources:

    Money Problems (p. 85)

    More Money Problems (p. 86)

    Division Problems (p. 87)

    One-Centimeter Graph Paper (p. 89)

    100 Chart (p. 90)

    300 Chart (p. 91)

Practice Pages (p. 93)

## Related Children's Literature

Dee, Ruby. *Two Ways to Count to Ten*. New York: Henry Holt, 1988.

Kasza, Keiko. *The Wolf's Chicken Stew*. New York: G. P. Putnam, 1987.

Silverstein, Shel. "Smart" in *Where the Sidewalk Ends*. New York: Harper and Row, 1974.

Viorst, Judith. *Alexander, Who Used to Be Rich Last Sunday*. New York: Atheneum, 1978.

An important part of students' mathematical work in the elementary grades is building an understanding of the base ten number system. This unit provides activities that develop knowledge about important *landmarks* in that system—numbers that are familiar landing places, that make for simple calculations, and to which other numbers can be related.

Because our number system is based on powers of ten, the numbers 10, 100, 1000, and their multiples are especially important landmarks. In solving real problems, people with well-developed number sense draw on their knowledge of these important landmarks. For example, think about how you would solve this problem, in your head, before you continue reading:

> If there are about 25 students in a class and 17 classes in our school, about how many students are there altogether?

Many people would use their knowledge that there are four 25's in every 100 to help them solve this problem mentally. Rather than multiplying 17 by 25, they will think something like this: "Four 25's in 100, eight in 200, 12, 16, that's 400, and one more 25 makes 425."

Knowledge about 10, 100, 1000, their multiples, and their factors is the basis of good number sense. As students learn about 100, how to take it apart into its factors, and how to use it to construct other numbers, they gain the knowledge they need to develop their own strategies to solve problems using quantities in the hundreds. They develop good estimation strategies and are less likely to make the kinds of errors that result from the use of faulty algorithms.

For example, a student who has developed knowledge about 20 and its relationship to 100, who has experience counting by 20's, and knows what the pattern of the multiples of 20 is like, would never make this common error:

$$
\begin{array}{r}
440 \\
- 380 \\
\hline
140
\end{array}
$$

Using a written subtraction algorithm—whether faulty or correct—is not a sensible approach to solving this problem. Rather, by inspecting the numbers and using knowledge of important landmarks in the number system, students should eventually be able to solve this problem mentally with no trouble:

> "380 to 400 is 20, then 20, 40, two more 20's gets to 440, so that's three 20's. The answer is 60."

**Mathematical Emphasis** At the beginning of each investigation, the Mathematical Emphasis section tells you what is most important for students to learn about during that investigation. Many of these mathematical understandings and processes are difficult and complex. Students gradually learn more and more about each idea over many years of schooling. Individual students will begin and end the unit with different levels of knowledge and skill, but all will gain greater knowledge about 100 and multiples of 100 and develop strategies for solving problems involving these important numbers.

Throughout the *Investigations* curriculum, there are many opportunities for ongoing daily assessment as you observe, listen to, and interact with students at work. In this unit, you will find two Teacher Checkpoints:

Investigation 1, Sessions 2–3:
Showing Factors of 36 or 48 (p. 12)

Investigation 1, Sessions 6–7:
Factors of 100 (p. 27)

This unit also has two embedded assessment activities:

Investigation 2, Sessions 1–3:
How Many in 500? (p. 36)

Investigation 3, Sessions 2–3:
Calculating with the 1000 Chart (p. 58)

In addition, you can use almost any activity in this unit to assess your students' needs and strengths. Listed below are questions to help you focus your observation in each investigation. You may want to keep track of your observations for each student to help you plan your curriculum and monitor students' growth. Suggestions for documenting student growth can be found in the section About Assessment (p. I–10).

## Investigation 1: Finding Factors

- How do students use cubes to count to 20? 36? 48? How do they arrange the cubes when they look for factors? How do they predict how many groups of a factor it will take to make 20? 36? 48? etc.? How do students make use of skip counting in looking for factors?

- Do students understand the relationship between factors and multiples? How do they transform a statement about factors to one about multiples?

- How do children show their fluency with skip counting? Are their strategies different when counting by two-digit numbers than one-digit numbers?

- What strategies do children use to predict how many 4's or 20's are in 100? How do they present their conclusions?

## Investigation 2: Using Landmarks to Solve Problems

- How do students figure out how many groups of a certain number are in a collection of 100 objects? How do they use this knowledge to make predictions about 200 and 300? How do they use this information to predict how many of that particular factor will be in higher numbers such as 500? In in-between numbers such as 640?

- How do students use information about factors and multiples in doing multiplication and division problems? Can they use what they know about 100 to figure out 500 divided by 25? Or 25 × 9?

- How do students use estimating strategies and knowledge of landmark numbers, factors, and multiples to help them find groups of equal size that total in the hundreds? Can they extend their strategies to 200's, 300's, and 400's?

- Do students recognize, interpret, and use standard forms and symbols for multiplication and division? Can they do this both on paper and when using the calculator? How do they interpret a problem that is presented in a variety of ways, including standard notation?

## Investigation 3: Constructing a 1000 Chart

- How do children figure out how many graph paper squares there are on their partially completed 1000 charts? How do they figure out how many more groups of 20's, 25's, or 100's they need to complete their chart? How do they locate multiples of 100 on their 1000 chart? Multiples of factors they have been using, such as 520, 375, 950?

- What strategies do students use to locate quantities of items in the classroom that are greater than 300? Around 1000? How do students find the corresponding numbers on the 1000 chart? What counting or grouping methods do they use to check to see if their estimate is accurate?

- How do students make use of landmark numbers when finding distances on the 1000 chart? Do they see the relationships between addition and subtraction and between multiplication and division?

In the *Investigations* curriculum, mathematical vocabulary is introduced naturally during the activities. We don't ask students to learn definitions of new terms; rather, they come to understand such words as *factor* or *area* or *symmetry* by hearing them used frequently in discussion as they investigate new concepts. This approach is compatible with current theories of second-language acquisition, which emphasize the use of new vocabulary in meaningful contexts while students are actively involved with objects, pictures, and physical movement.

Listed below are some key words used in this unit that will not be new to most English speakers at this age level, but may be unfamiliar to students with limited English proficiency. You will want to spend additional time working on these words with your students who are learning English. If your students are working with a second-language teacher, you might enlist your colleague's aid in familiarizing students with these words, before and during this unit. In the classroom, look for opportunities for students to hear and use these words. Activities you can use to present the words are given in the appendix, Vocabulary Support for Second-Language Learners (p. 65).

**the numbers 1 to 100**  Students use the 100 chart throughout the unit. They should be able to write the numerals and identify each by name.

**divide, group, equal, unequal**  As students learn about factors in this unit, they *divide* collections of cubes or amounts of money into *equal groups.*

**money: coins, cents, nickel, dime, quarter, dollar**  In order to "divide a dollar evenly among five people," students need to recognize U.S. coins and know the value of each.

**pattern**  Finding patterns is a key mathematical process. In this unit, students look for visual patterns on the 100 chart (for example, the diagonal pattern created when skip counting by 3's) and for number patterns (for example, when counting by 5's, alternating numbers end in 5 and 0).

## Multicultural Extensions for All Students

Whenever possible, encourage students to share words, objects, customs, or any aspects of daily life from their own cultures and backgrounds that are relevant to the activities in this unit. For example:

- Students who have coins from their countries of origin can bring them to show to the class. They might make a poster showing equivalencies of the coins in these monetary systems.

- When students are making up their own word problems, encourage them to write problems that are based on aspects of their cultures— foods, games and sports that involve teams, and so forth.

# Investigations

# Finding Factors

## What Happens

**Session 1: Skip Counting with Cubes**   Students are introduced to skip counting using interlocking cubes. By arranging 20 cubes in equal groups, they find and record "ways to make 20" (they are finding *factors*, but that term is not introduced until Session 2). Students also record groupings they try that "won't work to make 20." They use their cubes to find groupings (factors) for at least one other number.

**Sessions 2 and 3: Factors of 24, 36, and 48** These sessions begin with some work to introduce or review skip counting using a 100 chart. Then students are introduced to the term *factor* through discussion of their work with 20 in the previous session. Students find factors of 24, 36, and 48 and discuss what they notice about the factors of these numbers. Students record their findings by skip counting on a partial 100 chart and by using multiplication notation.

**Sessions 4 and 5: Factors of 100**   Students find factors of 100. They make a set of 100 charts for the numbers that are factors of one hundred, skip counting by one factor on each chart. They make a display showing 100 things, grouped by one of the factors of 100. They make a second display, grouped by a different factor, at home.

**Sessions 6 and 7: Dividing a Dollar**   The class lists and discusses all the factors of 100 they have found. Students use coins to explore how to share a dollar equally among five people, then investigate ways to divide a dollar evenly among different numbers of people.

## Mathematical Emphasis

■ Understanding the relationship between skip counting and grouping (for example, as we count 3, 6, 9, 12, we are adding a group of 3 to the total each time)

■ Becoming familiar with the relationships among commonly encountered factors and multiples (for example, is 3 a factor of 24? how many 3's does it take to make 24?)

■ Increasing fluency in counting by single-digit numbers (2's, 3's, 4's, 6's, 8's) as well as by useful two-digit numbers (10's, 20's, 25's)

■ Developing familiarity with the factors of 100, an important landmark in our number system, and their relationships to 100 (for example, that there are twenty 5's in 100 and five 20's in 100) through work with cubes, coins, and 100 charts

## What to Plan Ahead of Time

### Materials

- *Two Ways to Count to Ten*, retold by Ruby Dee (Holt, 1988) (Session 1, optional)
- Interlocking cubes: at least 50 per pair (Sessions 1–7)
- Overhead projector (Sessions 2–3, 6–7)
- Calculators (Sessions 4–5)
- Art materials for making pictures of 100: colored paper, scissors, glue or paste, crayons or markers, any scrap materials available (Sessions 4–5)
- Plastic or real coins: at least 50 pennies, 25 nickels, 12 dimes, and 6 quarters per small group of students. If using plastic, try to include at least one collection of real coins and let the groups take turns using them. Also, prepare another small sample of coins for each student pair: a few pennies, nickels, dimes, and one quarter, in differing total amounts from about 48 to 60 cents, in a plastic bag, envelope, or paper cup. (Sessions 6–7)

### Other Preparation

- Duplicate student sheets and teaching resources (located at the end of this unit) in the following quantities. If you have Student Activity Booklets, copy only the items marked with an asterisk, including any transparencies needed.

#### For Session 1
Student Sheet 1, Ways to Make 20 (p. 69): 2 per student

Student Sheet 2, Ways to Make _____ (p. 70): 4 per student (2 for homework)

Family letter* (p. 68): 1 per student. Remember to sign it before copying.

#### For Sessions 2–3
100 chart (p. 90): 2 per student (optional), and 1 overhead transparency*

Student Sheet 3, Factors of 24 (p. 71): 2 per pair, and 1 overhead transparency*

Student Sheet 4, Factors of 36 (p. 72): 3 per pair

Student Sheet 5, Factors of 48 (p. 73): 3 per pair

#### For Sessions 4–5
Student Sheet 6, Miniature 100 Charts (p. 74): 1 per student

Student Sheet 7, Factors of 100 (p. 75): 1 per student

Student Sheet 8, A Picture of 100 (p. 76): 1 per student (homework)

#### For Sessions 6–7
Student Sheet 9, Ways to Split Up a Dollar (p. 77): 1 per student

Student Sheet 10, Share a Dollar (p. 78): 1 per student (homework)

- If you plan to provide folders in which students will save their work for the entire unit, prepare these for distribution during Session 1.

# Skip Counting with Cubes

## Materials

- *Two Ways to Count to Ten* (optional)
- Interlocking cubes (20 per student or pair)
- Student Sheet 1 (2 per student)
- Student Sheet 2 (4 per student, 2 for homework)
- Family letter (1 per student)

## What Happens

Students are introduced to skip counting using interlocking cubes. By arranging 20 cubes in equal groups, they find and record "ways to make 20" (they are finding *factors*, but that term is not introduced until Session 2). Students also record groupings they try that "won't work to make 20." They use their cubes to find groupings (factors) for at least one other number. Their work focuses on:

- finding factors by making equal groups
- recording their work using pictures and numbers

## Activity

### Ways to Count to 10

If you have *Two Ways to Count to Ten* (a Liberian folktale), read the story as an introduction to skip counting. In this tale, the leopard holds a contest to choose his successor as king of the jungle. Each animal must throw a spear high up into the air. The animal who can count to ten before the spear reaches the ground will be the winner; he will be named prince and will marry the leopard's daughter. None of the animals succeeds until the clever antelope counts by 2's.

**Did the animals think of all the ways to count to 10? Are there any others?**

If you don't have the book, you might want to tell this story in your own words, or simply ask the class how many ways they can think of to count to ten:

**I know you can all count to ten—1, 2, 3, 4, 5, 6, 7, 8, 9, 10—but I'm thinking of a special way to count to 10. It's like a shortcut that gets us there faster. Who can think of a special way to count to 10?**

If students don't have any ideas, you might suggest:

**Is there a way I could do it starting with 2 instead of with 1?**

Count together by 2's. Can students think of any other ways to count to 10?

## Finding Factors with Cubes

**Building 20 with Groups of Cubes**  Give each student or pair of students 20 interlocking cubes and two copies of Student Sheet 1, Ways to Make 20. Students find as many ways as they can to build 20 using equal groups of cubes. For example, they might build 20 out of 2's:

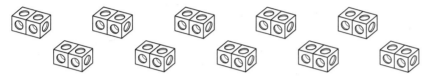

They record on Student Sheet 1 all the groupings they find that work and that don't work. For the groupings that do work, students both draw their cubes and record counting by that number, for example:

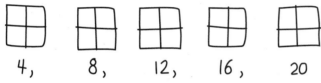

4,    8,    12,    16,    20

As you watch students work, make sure that they record all the groupings they try, even the ones that don't work. If a student announces "I'm finished," ask questions to prompt further thinking:

**Do you think you have all the ways? Why do you think so? Do you think 8 might work? How do you know?**

Name  Elena                                   Date  Jan. 5

**Student Sheet 1**

### Ways to Make 20

Take 20 cubes. Put them in equal groups.
How many different ways can you make 20?
Record your results here.

| Numbers I tried that will make 20 |
| --- |
| Number of cubes in group: __2__<br>Picture of 20:<br>Count: 2, 4, 6, 8, 10, 12, 14, 16, 18, 20 |
| Number of cubes in group: __5__<br>Picture of 20:<br>Count: 5, 10, 15, 20 |
| Number of cubes in group: _____ |

**Building More Numbers**   Distribute two copies of Student Sheet 2, Ways to Make ___, to each student. Have students choose another number to work with and write it in the blank in the title. Using the cubes, they find all the groupings that work to make their new number, just as they did for 20. Students can work in pairs or individually.

Help individual students choose a number appropriate for them. If 20 offered about the right level of challenge, those students could go on to 28 or 32. If 20 was difficult for them, they might try 12 or 16. Students who need a greater challenge might try larger numbers (40, 50, 60) or numbers with less familiar factors (21, 35, 42). **Note:** Don't use 24, 36, or 48, because students will work with these in later sessions.

As you circulate, ask students to show and tell you how they know a particular number "works." Encourage them to keep looking if they have not found all the possible groupings that work. Some students may not realize that they can count by numbers greater than 10, so remind them that they can.

## Session 1 Follow-Up

**Ways to Make _____**   Send home the family letter or *Investigations* at Home and two clean copies of Student Sheet 2. For homework, students find groupings that work for a number they have not yet done. Help students choose a number at an appropriate level of difficulty. Students record all the groupings they can find that work for their number. Briefly discuss with students objects they might use to count with at home: pennies, blocks, buttons, paper clips, pebbles, or drawings of dots on paper.

# Factors of 24, 36, and 48

## What Happens

These sessions begin with some work to introduce or review skip counting using a 100 chart. Then students are introduced to the term *factor* through discussion of their work with 20 in the previous session. Students find factors of 24, 36, and 48 and discuss what they notice about the factors of these numbers. Students record their findings by skip counting on a partial 100 chart and by using multiplication notation. Their work focuses on:

- finding factors by skip counting
- exploring the relationship between a multiple and its factors (how many 3's does it take to make 24?)
- developing facility in counting by 2's, 3's, 4's, 6's, and 8's
- making conjectures about factors and multiples (you can always use the number itself; 0 is never a factor; 2 is always a factor of any even number)
- using multiplication notation to record work with factors

**Ten-Minute Math: Counting Around the Class** Once or twice during the next few days, do Counting Around the Class. Remember, Ten-Minute Math activities are done outside of math time in any spare 10 minutes you have.

Choose a number to count by, let's say 3. Ask students to predict what number they'll land on if they count around the class exactly once. Encourage students to talk about how they could figure this out without doing the actual counting.

Then start the count: The first student says "3," the next "6," the next "9," and so forth. If students seem unfamiliar with what comes next, you may want to put numbers on the board as they count, so they can begin to see patterns.

Stop two or three times during the count to ask a question like this:

**We're at 33—how many students have counted so far?**

After counting around once, compare the actual ending number with their predictions.

For variations on this activity, see p. 61.

## Materials

- 100 chart (2 per student, optional)
- Transparency of 100 chart (optional)
- Interlocking cubes (50+ per pair)
- Transparency of Student Sheet 3
- Student Sheet 3 (2 per pair)
- Student Sheet 4 (3 per pair)
- Student Sheet 5 (3 per pair)
- Overhead projector

## Skip Counting on the 100 Chart

Consider the following to decide how much emphasis to give this activity:

■ If your students have *not* done skip counting on a 100 chart, they need to do this activity.

■ If your students did skip counting earlier in the year but could use some review and practice, you might try the activity with some less familiar numbers—6, 7, 8, or 9.

■ If your students have recently done a lot of skip counting on the 100 chart, you can go on to the next activity, What's a Factor? (p. 10).

**Counting by 2's**   Introduce or reacquaint your students with skip counting on the 100 chart. Showing the overhead transparency of the 100 chart, ask students to help you count by 2's. As the students tell you the next number in the series (2, 4, 6, 8, 10, 12, 14 ...), circle or put an X on it. When you get up to about 24, ask:

**As you're counting, how do you know what comes next?**

Encourage a variety of responses. Even though one student has already given a correct answer, ask other students for their way of thinking about it. Some students will begin to notice patterns on the 100 chart:

It's only the even numbers.
You just keep going under the ones you've already done.
You skip every other number.

Probably both you and your students will naturally use several different ways to refer to what you're doing on the 100 chart, such as "skip count by 2's," "count by 2's," "do the 2's pattern." This variety of language is valuable, as long as students understand that the phrases all refer to the same activity.

Distribute student copies of the 100 chart and have students complete the 2's pattern there. Circulate quickly and help students with any difficulties they may be having. Encourage them to double-check by comparing charts with each other. See the **Teacher Note**, Students' Problems with Skip Counting (p. 14); in particular, read about encouraging students to use cubes to build the groupings as they count.

Ask students to share the patterns they see.

**Counting by 3's**  Using a second copy of the 100 chart, students now count by 3's on their own. To help them get started, you might have them tell you the first few numbers of the 3's pattern, up to about 15, while you mark them on a clean transparency.

While students are working on their 3's, fill in the transparency by 3's up to 30. When students are finished with their own charts, ask:

**What are some ways you can tell what number comes next on my chart?**

| 1 | 2 | 3 | 4 | 5 | 6 | 7 | 8 | 9 | 10 |
|---|---|---|---|---|---|---|---|---|-----|
| 11 | 12 | 13 | 14 | 15 | 16 | 17 | 18 | 19 | 20 |
| 21 | 22 | 23 | 24 | 25 | 26 | 27 | 28 | 29 | 30 |
| 31 | 32 | 33 | 34 | 35 | 36 | 37 | 38 | 39 | 40 |
| 41 | 42 | 43 | 44 | 45 | 46 | 47 | 48 | 49 | 50 |
| 51 | 52 | 53 | 54 | 55 | 56 | 57 | 58 | 59 | 60 |
| 61 | 62 | 63 | 64 | 65 | 66 | 67 | 68 | 69 | 70 |
| 71 | 72 | 73 | 74 | 75 | 76 | 77 | 78 | 79 | 80 |
| 81 | 82 | 83 | 84 | 85 | 86 | 87 | 88 | 89 | 90 |
| 91 | 92 | 93 | 94 | 95 | 96 | 97 | 98 | 99 | 100 |

Responses might include:

> Count three more from 30.
> It's the next one in the diagonal, 6, 15, 24, 33.
> It's just like the first line—it's going to go 3, 6, 9 again.

Keep asking until there are no more suggestions.

Now fill in your own chart up to 63. Ask the students what ways they can think of to predict the next number.

When the counting is completed, ask students what patterns they can see in counting by 3's. Let them point out patterns on the transparency. Discuss further:

**How is the 3's pattern different from the 2's pattern?**

# What's a Factor?

Make a class list of all the numbers that make equal groupings for 20 (those that the class discovered in Session 1). These are the *factors* of 20. They include 1, 2, 4, 5, 10, and 20. Also list numbers students tried that did not work for 20. Pose questions for discussion:

**How do you know that 3 doesn't work? How did you find out that 20 works? Were there any numbers that surprised you? Do you think we have all the factors of 20? Could there be any others? Why do you think so?**

In this discussion, begin to use the word *factor*, but don't insist that students use this word. See the **Teacher Note**, Introducing Mathematical Vocabulary (p. 15).

❖ **Tip for the Linguistically Diverse Classroom** Instead of asking students to articulate their answers, ask them to communicate their thoughts by using visual aids, either cubes or marks on the board. For example:

**Show that 3 is NOT [*shake head*] a factor of 20.**

Student groups 20 cubes or marks on the board as follows:

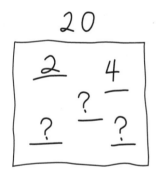

As students agree on numbers that are factors of 20, write them inside a square with the number 20 written above it. Add question marks with lines beneath them inside the square. Pointing to these lines, ask if there might be more factors for [*point*] 20. If students think so, have them come forward, erase the question marks, and write these numbers on the blank lines. Challenge them to prove that these numbers are indeed factors.

# Making Counting Charts for 24

Each pair of students needs a supply of cubes and two copies of Student Sheet 3, Factors of 24.

**We're going to try to find all the factors of 24. What's a number that we can count by to land exactly on 24?**

Take a student suggestion. Ask students to count by that number to 24. Using the overhead transparency of Student Sheet 3, mark the numbers as they count. Then focus on how many of that number it would take to make 24. For example, if they counted by 2's, ask:

**If you built 24 out of 2's, how many 2's would you need?**

Ask students to prove their answer by using the cubes. As students work, circulate quickly to see if most can show that twelve 2's make 24.

Finally, ask students how they could write this result using a multiplication sentence. If they don't know, show them this sentence:

$12 \times 2 = 24$

Explain that it can be read as "12 groups of 2 makes 24."

See the **Teacher Note**, What About Notation? (p. 21), for more information on helping students understand the connection between the relationships they are showing with cubes or counting charts and standard multiplication and division notation.

Next introduce the student sheet activity:

**Mathematicians think of 24 as a special number because it has a lot of factors. You're going to see how many factors of 24 you can find. For each factor, use one of the counting charts on Student Sheet 3. Show how you count by that number on the chart, then fill in the sentences that tell about that factor.**

Students work in pairs to find the factors of 24. Since there are eight factors of 24 (1, 2, 3, 4, 6, 8, 12, 24), each pair will need two copies of Student Sheet 3.

I found a secret! If you count by 24, you land right on 24!

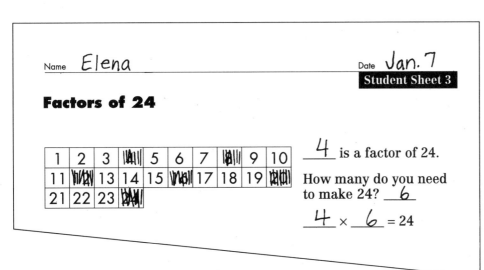

Name __Elena__                     Date __Jan. 7__
Student Sheet 3

**Factors of 24**

| 1 | 2 | 3 | 4̶ | 5 | 6 | 7 | 8̶ | 9 | 10 |
|---|---|---|---|---|---|---|---|---|----|
| 11 | 1̶2̶ | 13 | 14 | 15 | 1̶6̶ | 17 | 18 | 19 | 2̶0̶ |
| 21 | 22 | 23 | 2̶4̶ | | | | | | |

__4__ is a factor of 24.

How many do you need to make 24? __6__

__4__ × __6__ = 24

## Factors of 36 and 48

**Making Charts for 36 or 48**  Distribute three copies of Student Sheet 4, Factors of 36, and Student Sheet 5, Factors of 48, to each student pair. Students continue as in the preceding activity, marking charts for the factors of either 36 or 48. Many students will work on just one of these numbers; some students might want to work on both. You might assign one of ⟨the⟩ ⟨n⟩umbers for homework, in which case students will need their own ⟨copie⟩s of the student sheet.

⟨Encou⟩rage discussion within small groups of students working near each ⟨other.⟩ What are they finding out about 24, 36, and 48? Are there similari⟨ties/di⟩fferences? Note any interesting observations and encourage students ⟨to contr⟩ibute these later to the whole-class discussion.

## Factors of 36 or 48

⟨At some po⟩int while students are working on their 36 and 48 charts, stop ⟨the c⟩lass and ask everyone to do the following:

**Use your cubes to show me one of the ways to count to either 36 or to 48. You should arrange your cubes in such a way that I can tell, just by looking at your cubes, what factor you chose and whether it makes 36 or 48.**

Give students a few minutes to arrange their cubes. Once the cubes are arranged, students can leave them in place and go back to their work on the 36 and 48 charts while you circulate. As you walk around to take a quick look at each student's work, your role is to be the audience:

**Let's see if I can tell how you counted by looking at what each of you did with the cubes.**

You can also ask individual students how many groups they needed to make the total:

**I see you used 4's to make 36. Do you know how many 4's you needed?**

This brief check will give you a chance to reflect on how comfortable the class as a whole is becoming with finding factors and relating them to the multiple with which they are working. Do students understand that when they skip count by a number, they are accumulating groups of that number of objects? Can they skip count easily using the grouping they have chosen?

## Class Discussion: What Did You Find Out About 24, 36, and 48?

When students have worked on at least two of these numbers, have a brief class discussion in which students describe what they have noticed about the factors of these numbers. You don't need to wait until students are finished with all the student sheets in order to have this discussion.

This is a good time to share observations that you have heard made by individual students or in small groups.

❖ **Tip for the Linguistically Diverse Classroom** Include discussion questions that can be simply answered by a hand-showing or yes/no response. For example:

> Raise your hand if you think 24 and 36 share some of the same factors.
>
> Is 2 a factor of both 24 and 36?

Fill squares with the factors of 24 and 36. Ask students to point to the factors in the squares that are the same.

At this point, students will probably come up with some general conjectures about factors. You may want to make a class list of these to keep (see the **Dialogue Box**, Factors of 24, 36, and 48, on p. 16). List any ideas that students feel they have good evidence for now, even though the ideas may need to be tested further. You can return periodically to these conjectures throughout the unit.

---

## Sessions 2 and 3 Follow-Up

**Factors of 36/Factors of 48** Students can take home Student Sheet 4, Factors of 36, or Student Sheet 5, Factors of 48, to complete as homework if they do not finish in class. Remember that students will need three copies of each sheet.

🏠 **Homework**

Some students have difficulty keeping track of their skip counting on the 100 chart. Here are some confusions we have noticed in classrooms:

■ Some students always start on 1, no matter which number they are skip counting by.

■ The count may get off by 1 because the student pauses at a circled number, then starts counting again with that number. For example, when counting by 6's, a student counts 6, 12, 18, then begins the next count on 18. After counting six more numbers (18, 19, 20, 21, 22, 23), the student lands on 23 instead of 24.

■ Students sometimes follow a "false pattern" that doesn't actually work for the number they are counting by. For example, they may circle 3, 6, 9, then color straight down the columns under the 3, 6, and 9, not realizing that the 3's pattern doesn't continue in columns the way the 2's pattern does.

■ Students may miscount one interval and then continue counting correctly, so that all subsequent numbers are affected by the original mistake. For example: 3, 6, 9, 12, 15, 19, 22, 25, 28 ....

Some of these difficulties are simply miscounting mistakes that anyone can make. Help students to use the pattern on their counting charts to check: Does the pattern continue consistently on the chart? Also, have students double-check each other. When two or three students compare charts, they can often find and correct their own miscounting.

However, some students may truly not understand what they are doing when they "count by 2's" or "count by 3's" on their charts. Here, using cubes as a first step will help. That is, when counting by 2's, the student makes a group of 2 cubes, then marks 2 on the chart; makes another group of 2 cubes (perhaps in a different color), and marks the total, 4; then makes another group of cubes, marks the total, 6; and so forth. Students will naturally stop using the cubes as soon as they feel comfortable with skip counting.

We have found that it's not helpful for students to use cubes to mark squares directly on the counting charts. Students can't see the numbers underneath them, and they often move a cube accidentally to a neighboring square, thereby misleading themselves about the pattern on the chart.

# *Introducing Mathematical Vocabulary*

In this unit, several important mathematical words come up naturally in discussing the activities. Introduce these words by beginning to use them yourself. Explain what you mean by them, but don't insist that students use them.

**Factor** Ways of referring to the numbers that "work" in these activities can be wordy or cumbersome—"numbers you can count by to land exactly on 20," or "equal groupings that make 20." You can gradually introduce the word *factor*. If the introduction of a vocabulary word is preceded by activities that make its definition clear, students enjoy knowing an "adult" word to refer to a concept they have learned.

**Even and Odd** These words will come up in the students' descriptions of patterns on the 100 chart. Don't assume that students know exactly what they mean by these words. When one teacher asked, "What do you mean by *even?*" to a child who was describing the pattern of 2's on the 100 chart, he provoked a 45-minute discussion about the characteristics of odd and even! Some children believe that an even number has only even factors: "No, 3 isn't a factor of 24 because 3 isn't even." This is a good conjecture to have students investigate.

**Row and Column** In talking about their work with the 100 chart, students often confuse the words *row* and *column*, describing a pattern as going "down the row" rather than "down the column." This is a good opportunity to talk about the difference, since using *row* for both (as students often do) makes communication more difficult. However, do not insist that students use these words in the conventional way as long as they can explain or demonstrate what they mean. Remembering the difference can be hard, and focusing on getting the words right may obscure the good mathematical thinking a student is doing. Rather, keep using the terms yourself so that students continually hear them used correctly in context. Other terms that may come up in this context and may need some explanation are *horizontal*, *vertical*, and *diagonal*.

**Multiple** You may want to hold off on this word until *factor* is well established in the classroom vocabulary, but it can be naturally introduced in Investigation 2, when students are dealing with multiples of 100: "Maya noticed that when we're counting by 25's, after every four numbers we get to a *multiple* of 100."

columns

↓

rows →

| 1 | 2 | 3 | 4 | 5 | 6 | 7 | 8 | 9 | 10 |
|---|---|---|---|---|---|---|---|---|----|
| 11 | 12 | 13 | 14 | 15 | 16 | 17 | 18 | 19 | 20 |
| 21 | 22 | 23 | 24 | 25 | 26 | 27 | 28 | 29 | 30 |
| 31 | 32 | 33 | 34 | 35 | 36 | 37 | 38 | 39 | 40 |
| 41 | 42 | 43 | 44 | 45 | 46 | 47 | 48 | 49 | 50 |
| 51 | 52 | 53 | 54 | 55 | 56 | 57 | 58 | 59 | 60 |
| 61 | 62 | 63 | 64 | 65 | 66 | 67 | 68 | 69 | 70 |
| 71 | 72 | 73 | 74 | 75 | 76 | 77 | 78 | 79 | 80 |
| 81 | 82 | 83 | 84 | 85 | 86 | 87 | 88 | 89 | 90 |
| 91 | 92 | 93 | 94 | 95 | 96 | 97 | 98 | 99 | 100 |

# Factors of 24, 36, and 48

In this discussion, related to the discussion activity on p. 13, students are sharing the things they've discovered about the factors of 24, 36, and 48.

**Saloni:** I found a secret! If you do 24, you land right on 24!

**Does that work for other numbers?**

**Saloni:** Yes!

**Annie:** The number always works for itself, like 36 works for 36, and 48 works for 48.

**Sean:** The 3's worked on all of them.

**Tamara:** The 6's work for all of them, and the 2's work for all of them.

**Kate:** It's because they're all even numbers.

**Aaron:** Yeah, 2 will always work if it's even, 'cause counting by 2's is the even numbers.

**Maria:** If you can make 6's then you can make 3's, because I was doing it with the cubes, and I had 6's for 24, so then I just cut them in half, and then I had 3's.

**Sean:** And 9 always works when you have 3's.

**Kate:** I counted by 1's and it works on all of them.

**Ryan:** I agree with Kate. It doesn't matter what number, you can always count it by 1's.

**Latisha:** Wait, I don't think it's right what Sean said about the 9's. Nine doesn't work for 24.

**Sean:** It has to, because 3 does.

**Latisha:** No, because 9 and 9 is 18 and then another 9 would be too many.

**Ricardo:** Zero doesn't work.

**Why doesn't that work?**

**Ricardo:** Because you don't get anywhere!

From this discussion, the teacher recorded the following conjectures, which the class returned to throughout the unit as students worked with other factors and multiples:

- The number always works for itself.
- 2 will always work if it's even.
- If 6 is a factor, then 3 is a factor too.
- 1 is a factor of every number.
- 0 is never a factor.

Some teachers like to identify particular children as the authors of conjectures; for example, "Kate's conjecture—1 is a factor of every number."

---

❖ **Tip for the Linguistically Diverse Classroom**
Following the pattern established in the tip on p. 10, include visual diagrams to support each conjecture. For example:

- 1 is a factor of every number.

- If 6 is a factor, then so is 3.

Provide opportunities for students who cannot articulate their conjectures to come to the board to present their ideas using similar visual illustrations.

---

# Factors of 100

## What Happens

Students find factors of 100. They make a set of 100 charts for the numbers that are factors of one hundred, skip counting by one factor on each chart. They make a design showing 100 things, grouped by one of the factors of 100. They make a second design, grouped by a different factor, at home. Their work focuses on:

- finding which numbers are factors of 100
- skip counting with the factors of 100
- figuring out how many of a certain factor make 100 (for example, that four 25's make 100)

 **Ten-Minute Math: Counting Around the Class**  During the next few days, continue to do Counting Around the Class in short sessions outside of the mathematics hour.

Count by 4's or 5's. Before you begin, ask:

**Do you think our final number will be more than 100? Why do you think so?**

Stop two or three times during the count and ask questions like this:

**We're at 65 now—how many more students will have to have turns to reach 100?**

Some days, you might have students use a calculator for this activity. On most calculators, the equals (=) key allows you to skip count easily. For example, to skip count by 5's, press a starting number (typically, 0), the operation (in this case, +), and the number you want to count by (in this case, 5). Then, press the equals key each time you want to add 5. Thus, when you press

you will see on your screen 5, 10, 15, 20.

For full directions and variations on this activity, see p. 61.

## Materials

- Interlocking cubes
- Colored paper, scissors, glue or paste, crayons or markers, scrap materials
- Student Sheet 6 (1 per student)
- Student Sheet 7 (1 per student)
- Student Sheet 8 (1 per student, homework)
- Calculators

# What Numbers Are Factors of 100?

**Introducing "Landmark" Numbers** Briefly discuss the general idea of landmarks:

**What is a landmark? Why are they useful? Have you ever used landmarks, or seen anyone in your family use landmarks? Can a tree be a landmark? How about a building? Or a sign for a business?**

Explain that there are important numbers that we can use as "landmarks," to help us tell where we are when we are counting or figuring.

**When you count, do you get the feeling that some numbers are more important than others? If so, which ones? Why are these important?**

Students might suggest 2, 5, 10, 50, 100, or others. Explore their sense of these.

**When we reach a number like 10, or 50, or 100, it makes us feel comfortable—we know where we are. That's why we can think of numbers like these as "landmark numbers." The number 100 is a particularly important landmark in our number system. We're going to spend some time learning all about 100, as well as the other hundreds, which are all landmark numbers: 200, 300, 400, 500, 600, 700, 800, 900, 1000.**

**Ways to Make 100** Students work in pairs or small groups to find all the factors of 100. Give each student a copy of Student Sheet 6, Miniature 100 Charts, and Student Sheet 7, Factors of 100. (Although they work collaboratively, each student does his or her own recording.) For each factor of 100, a student marks a miniature 100 chart, counting by that number and recording how many "jumps" it took to reach 100. Interlocking cubes remain available for students who want to make equal groups to test whether a number is a factor of 100. (Students who have found factors of 100 in the Multiplication and Division unit, *Things That Come in Groups*, may move through this activity quickly.)

Let's try 65. Do you think it will work?

Well, no, because 65 plus 65 is more than 100— but let's try it to see.

Students use Student Sheet 7 to keep a list of every number they try, whether it does or does not "work" as a factor of 100. They also figure out how many of each factor makes 100 and record this information on Student Sheet 7. Students keep this work in their folders for reference throughout the unit. The miniature 100 charts might be cut out from Student Sheet 6 and made into a small book.

As you observe students working, pose questions for discussion:

**What did you find out so far? How do you know 3 doesn't work? How did you find out that 20 works? Were there any numbers that surprised you? Do you think you have all the factors of 100? Could there be any others? Why do you think so?**

❖ **Tip for the Linguistically Diverse Classroom** Ask students to communicate their thoughts through visual aids, as explained in the tip on p. 16.

You will be following up on this activity in the next session by helping students make a class chart showing all the factors of 100 they found (see chart, p. 23).

**Activity**

Students each make an arrangement that shows 100 of the same thing, coloring or pasting objects onto paper. Their 100 things should be arranged in groups—one of the factors of 100—so that someone looking at the picture could easily count the objects. For example, students might arrange their objects by 5's or by 10's—but let them choose their own groupings.

# Making a Picture of 100

If students are drawing, encourage them to keep their pictures simple since they will have to draw a lot of them! Lay out any art or scrap materials on a table. Encourage students to be creative, using anything available in the classroom.

- Students have drawn pictures of 100 hearts or 100 balloons.
- Some students have used rubber stamps.
- One student made rubbings of the bumps on plastic building blocks.
- Another student made a collection of tiny strips of paper that she painstakingly cut out.
- One student wrote a list of 100 names.
- Another used a hole puncher to punch out 100 "holes" from colored paper, which he then glued onto another sheet.

**Homework**

**A Picture of 100** Students create a second arrangement of 100 at home, using a different factor. Directions are on Student Sheet 8, A Picture of 100. You may also want to send home sturdy, oversize paper for them to work on. Families are encouraged to participate in making the display with their child.

Avoid specific suggestions for what to use for the home chart; emphasize that students may use anything they want. Some students have used buttons, macaroni, paper clips, soda can pop-tops, popcorn kernels, toothpicks, dried beans, and rice grains. They can also draw pictures.

❖ **Tip for the Linguistically Diverse Classroom** Give students the chance to circulate around the class and look at the pictures of 100 their classmates made during the session, to help them understand the variations that are possible.

You might want to make a bulletin board display of the students' collections, with a heading something like "100 of Each."

**Note:** The class will discuss this work during the first session of Investigation 2; be sure to save their pictures for that discussion.

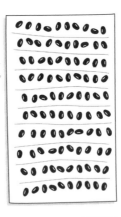

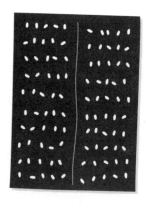

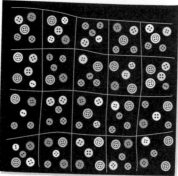

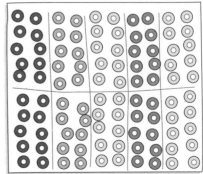

# What About Notation?

It is important that your students learn to recognize, interpret, and use the standard forms and symbols for multiplication and division, both on paper and on the calculator. In this unit, students will use these:

$$12 \qquad 3 \times 12 \qquad 12 \div 3 \qquad 3\overline{)12}$$
$$\underline{\times 3}$$

Your challenge is to introduce these symbols in a way that allows students to interpret them meaningfully. That is, students must understand *what is being asked* in a problem that is written in standard notation. They can then devise their own way to find an answer. Notation is also useful as an efficient way to record a problem and its solution. It is not just a directive to carry out a particular procedure, or a signal to forget everything you ever knew about the relationships of the numbers in the problem!

Your students may come to you already believing that when they see a problem like $3\overline{)42}$, written in the familiar division format, they must carry out the traditional long-division procedure. Instead, we want them to use everything they know about these two numbers in order to solve the problem. They might skip count by 3's out loud or on the calculator. (For tips on skip counting on the calculator, see the Ten-Minute Math activity, p. 17.) Or they might use reasoning based on their understanding of number relationships:

> It takes ten 3's to make 30. Then there are three more 3's to get up to 39, that's thirteen 3's so far. Then 40, 41, 42—that's one more 3—it's 14!

> Well, half of 42 is 21, and I can divide 21 into seven groups of 3, so you double that, and it's 14.

Similarly, when students see a multiplication problem like $4 \times 55$ written vertically, they are likely to forget everything they know about these numbers and try to carry out multiplication with carrying. Instead, we want students to use what they know about landmarks in the number system and other familiar number relationships. For example:

> I know that two 50's make 100, and there's four 50's, so that's 200. Then I know that four 5's is 20, so it's 220.

Students need to get used to interpreting multiplication in both horizontal and vertical form as simply indicating a multiplication situation, not a particular way to carry out the problem. So, while you help students to read standard notation and to use it to record their work, keep the emphasis on understanding the problem context and using good number sense to solve the problem.

# Dividing a Dollar

## Materials

- Interlocking cubes
- 100 charts
- Sample coin collections in differing total amounts (48 to 60 cents) for each pair
- A container of coins for each small group (at least 50 pennies, 25 nickels, 12 dimes, 6 quarters)
- Student Sheet 9 (1 per student)
- Student Sheet 10 (1 per student, homework)
- Overhead projector

## What Happens

The class lists and discusses all the factors of 100 they have found. Students use coins to explore how to share a dollar equally among five people, then investigate ways to divide a dollar evenly among different numbers of people. Their work focuses on:

- continuing to explore which numbers are and are not factors of 100, using money as a context
- continuing to become familiar with how many of a certain factor make 100 (for example, that four 25's make 100), in the context of money
- determining the value of a collection of coins

 **Ten-Minute Math: Counting Around the Class** In short sessions at various times during the day, continue to do Counting Around the Class as described in Sessions 4 and 5. Try counting by 10's. For other variations, see p. 61.

## Activity

### Listing the Factors of 100

Work with students to make a large class chart of all the factors of 100, based on their work in the preceding sessions. You'll need to make a chart that can remain posted. Have students help you list these factors from the least to the greatest. Ask:

**Do you think we have all the factors of 100? Why do you think so? Are there any numbers you didn't try that might work?**

Then add to the chart how many of each factor it takes to make 100, asking students to supply these numbers:

**How many 20's make 100? How do you know? Who can prove it a different way? How many 2's make 100? How do you know? Who has a different way to explain the solution?**

Encourage students to refer to their work with cubes and 100 charts as ways to explain what they think. For example:

> When I did it with cubes, we did it in 20's, and we needed 5 rows of 20's.

> I can tell it's five 20's because I counted on the 100 chart—20, 40, 60, 80, 100, that's five.

Add these results, with the associated words and multiplication equations, to your chart. Include only the factors your students have found.

| Factor of 100 | How many make 100? | Say it in words | Say it as an equation |
|---|---|---|---|
| 1 | 100 | 100 groups of 1 | $100 \times 1 = 100$ |
| 2 | 50 | 50 groups of 2 | $50 \times 2 = 100$ |
| 4 | 25 | 25 groups of 4 | $25 \times 4 = 100$ |
| 5 | 20 | 20 groups of 5 | $20 \times 5 = 100$ |
| 10 | 10 | 10 groups of 10 | $10 \times 10 = 100$ |
| 20 | 5 | 5 groups of 20 | $5 \times 20 = 100$ |
| 25 | 4 | 4 groups of 25 | $4 \times 25 = 100$ |
| 50 | 2 | 2 groups of 50 | $2 \times 50 = 100$ |
| 100 | 1 | 1 group of 100 | $1 \times 100 = 100$ |

Ask students what they notice about the list. Someone may notice related pairs; for example, they may see 20 groups of 5 and 5 groups of 20 (or, two 50's and fifty 2's). If no student notices this, you might point out one of the matching pairs yourself. This discussion can lead students to identify factors that are not yet on their list. Those students who have already worked with arrays in the Multiplication and Division unit, *Things That Come in Groups*, may be able to describe why they think both pairs work.

If the chart is not yet complete, or if your students are not sure whether they have all the factors, tell them that you will keep the chart posted and add to it if they find more factors. You may also want to post miniature 100 charts showing the pattern for each factor.

## Working with Coins

Give pairs of students the small sample coin collections (a few pennies, nickels, dimes, and at least one quarter, in differing amounts from about 48 to 60 cents). Review the values of coins with your class:

**What is a nickel worth? A dime? A dime and a nickel together? How do you know? What is a quarter plus a nickel worth? How do you know? Which is the biggest coin? How much is it worth? Is the smallest coin worth the smallest amount of money?**

............................................................................................................................

❖   **Tip for the Linguistically Diverse Classroom**   Reword questions so that students can respond nonverbally. For example:

> **Show me a coin that is worth 5 cents.**
>
> **Show me a coin that is worth 10 cents.**
>
> **Show me two coins that equal 30 cents.**

............................................................................................................................

Ask each pair to figure out how much money they have altogether. You may want to have each pair trade its coin collection with another pair and calculate the new total.

Have one pair tell how many of each coin is in their collection. Draw these on the board or put them on the overhead. Ask the class to come up with different counting strategies for figuring out the total.

**Note:** Each day you use money, students enjoy finding out how much their collection is worth before doing other activities. This provides for good practice in counting by 5's, 10's, and 25's. It also allows students to check their collection at the end of class to make sure they have the same total before putting their coins away.

## Dividing a Dollar Among Five People

**We're going to be finding out more about 100 by using money. If I had a dollar in pennies, how many pennies would I have?**

Establish that there are 100 pennies or 100 cents in a dollar.

**We're going to work on how we can split up 100 pennies evenly. If I had a dollar, could I share it equally between Midori and Chantelle? How much would each person get?**

Students will probably know how to divide a dollar in half. Hold a short discussion of how much each person would get and how the students know that. What coins could they use to make this amount?

Give each small group their container of additional coins, and pose the following problem for work in pairs or small groups:

**Now, could I split a dollar equally among Midori, Chantelle, Sean, Michael, and Khanh? How much would each person get? Find as many different ways as possible that you could divide a dollar equally among five people.**

In this activity, giving two dimes to each person is considered different from giving four nickels to each person. You can clarify this as you circulate. Once students have one solution, you can explain what a "different solution" is by saying something like this:

**Suppose you didn't have enough dimes to give everyone two dimes. Could you use some different coins to give each person an equal amount?**

**Recording and Sharing Solutions** Students work in small groups, recording each way they find so that they can remember and report their solutions later. They might use pictures, words, symbols, or a combination of these. We have deliberately not provided a student sheet for recording their findings, so that students can devise their own way of recording their solutions. As you circulate, remind students to record each solution with words, numbers, or pictures.

Try to interpret their recording. If it is unclear, insist that they clarify:

**I can tell that you used nickels, but I don't understand how many nickels each person would get.**

After a while, have groups share their answers with the whole class. Ask each group to share a *new* method—one that has not yet been shown by someone else. To share their solution, they might specify in a sentence what each person gets, or they might arrange the coins on the overhead projector. You might want to keep a chart of their findings.

Your students may not realize at first that every solution they find involves giving 20 cents to each person. The **Dialogue Box**, Dividing a Dollar Among Five People (p. 28), illustrates how one pair discovered that they always had to use 20 cents.

## How Can We Divide a Dollar Evenly?

5/20

| Number of people sharing a dollar | How much each gets |
|---|---|
| 2 people | 50¢ |
| 5 people | 20¢ |
| Cannot split a dollar evenly 3 people | |

**Dividing a Dollar Among Three People**   Pose the following question to the whole class:

**Could you share a dollar equally among three people—Latisha, Mark, and Samir? How could you do it?**

There is a way, but it wouldn't be a fair way. Someone would have 34 cents and the rest 33 cents.

After sharing a dollar among two and five people, students may think that it is possible to split a dollar among any number of people. Trying to split a dollar among three people, however, will be puzzling to many.

One approach to this discussion is to put 10 towers of 10 cubes each on the overhead. Ask students what they would do to divide these "pennies" up among three people. See the **Dialogue Box**, 100 Cents and 3 People (p. 29), for part of a discussion that occurred in one classroom.

If this problem remains puzzling to some, leave it as an open question for students to continue exploring, working with coins or cubes in their groups.

**How Can You Split a Dollar Evenly?**   Review the class findings so far about how a dollar can be divided evenly. Start a class chart, as shown. Include numbers of people the students know cannot divide a dollar evenly.

Students may know some other numbers they can add to the chart; for example, that 10 people would each get 10 cents, or that 99 people cannot divide a dollar evenly. If students give such information, add it to the chart.

Students then work in pairs to find out what other numbers of people could share a dollar equally. Hand out Student Sheet 9, Ways to Split Up a Dollar, for their recording. To get students started, suggest that they choose one or two of the following to explore: 4 people, 8 people, 25 people, 50 people, or 30 people.

They may also choose other numbers that interest them. Some students will work on just one or two of these problems. Others may become interested in finding all the possible ways to divide a dollar evenly. In either case, it's important that each student be able to describe and justify his or her solutions.

Some students may come up with solutions that don't involve whole cents, for example, giving each of three people 33 1/3 cents. This is a good opportunity to talk about fractions of wholes and for the class to see and interpret fraction notation. Acknowledge that such an answer is an interesting mathematical solution, but that it would be impossible to actually give anyone 1/3 of a cent. You may also want to make a separate category on your class list: "Solutions that work with numbers, but not with pennies."

**Summing Up**  Add student findings to your class chart.

| Number of people sharing a dollar | How much each gets |
|---|---|
| 2 people | 50¢ |
| 4 people | 25¢ |
| 5 people | 20¢ |
| 10 people | 10¢ |
| 20 people | 5¢ |
| 100 people | 1¢ |

Cannot split a dollar evenly
3 people
8 people
anything over 50

## Activity

## Teacher Checkpoint
# Factors of 100

By this time, students should be quite familiar with the factors of 100. This is a good time to pause and make sure that your students are comfortable with the relationships between 100 and its factors, before you go on to work with higher numbers. Pose these two questions (write them on the board or overhead):

> **How many 20's are in 100?**
> **How many 4's are in 100?**

Ask students to find the answers to these and to prove their solutions using coins, cubes, or 100 charts, then write or draw explanations of their solutions. The purpose of their writing and drawing is to demonstrate convincingly that their solution is correct. If you are keeping student portfolios, this piece of work could be one to save.

> There are 25 4's in 100.
>
> I counted by 4 on the 100 chart.
>
> I found 25 4's in a zig zag line.

There may be some students who need more work with 100. When you go on to work with multiples of 100 in Investigation 2, you can adjust the numbers for those students. Thus, instead of finding how many 25's in 200 and 300, they might find how many 10's in 70, 80, and 90; how many 20's in 40, 60, and 80; and so forth.

## Sessions 6 and 7 Follow-Up

**Share a Dollar**  Send home Student Sheet 10, Share a Dollar. Students try to find a way to share $1.00 among 10 people. Then they make up their own share-a-dollar problem and solve it. You may want to allow time for them to solve one another's problems during class.

 **Homework**

## Dividing a Dollar Among Five People

This discussion between two students takes place while they are working together on the activity Dividing a Dollar Among Five People (p. 24).

**See how you could share a dollar equally among five people. I want you to find as many combinations as possible.**

[*Yoshi and Amanda are working together. Amanda lines up 5 dimes.*]

**Yoshi:** We found one—two dimes!

**Amanda:** Wait, I'm not sure yet. [*She adds a second dime on top of each of the first 5.*]

**Yoshi** [*impatiently*]: That's it!

**Amanda** [*not yet satisfied*]: We need to check it. Fifty plus fifty … yes.

**Yoshi** [*already onto another solution*]: They could each get 4 nickels … it's the same thing!

[*Amanda begins to try nickels, making piles of 4.*]

**Amanda:** I don't know if we've got enough to do it.

**Yoshi:** Yes, 4 nickels is the same thing. And we could use 20 pennies, it's the same thing. It works.

[*Amanda, working on a solution independently, places 1 dime and 2 nickels in each of five piles.*]

**Amanda** [*counting the dimes first*]: 10, 20, 30, 40, 50 … [*then the nickels*] 55, 60, 65, 70, 75, 80, 85, 90, 95, 100. [*She then tries groups of 5 nickels.*]

**Yoshi:** That's too many.

**Amanda:** No, wait, I want to try it! Don't worry, I know what I'm doing.

[*Amanda counts her nickels by fives. After the fourth pile, her count is already up to $1.00, with 5 nickels left, so she abandons this idea.*]

**Amanda:** Let's think of another one not using 20 cents. [*Pause.*] No, we have to use 20 cents, don't we?

**Yoshi:** We could do 10 pennies and 2 nickels, that would work.

[*Amanda again gets the coins to try it, as she is not convinced without using the coins.*]

## 100 Cents and 3 People

This class discussion takes place during the activity How Can We Divide a Dollar Evenly? (p. 26).

**Could you share a dollar equally among three people? How would you do it?**

**Elena:** 30 pennies.

**Dylan:** That equals 90.

[*The teacher puts three groups of 30 on the overhead.*]

**How about these 10 left over?**

**Ricardo:** I think you should put them in 20's.

[*The teacher puts out three groups of 20.*]

**Ricardo:** Give 10 to the 20's.

**Su-Mei:** It still equals 90.

**Aaron:** I don't see any way … I don't think you could do it.

**Yoshi:** 34.

**So keep these 30's and add 4 more to each? What do you think is going to happen then?**

**Latisha:** Two cents will get left over. You'll get 4 and 4 each and that's 8, and 8 from 10 is 2, so there's 2 left over.

**Sean:** I think 2's would work. No, do it by 1's.

**Jamal:** I think if you gave 33 each, you'd have one left over and you could divide it into thirds, if you had a butcher knife or something.

**Tamara:** It's impossible to divide 10 into three because 10 is an even number and 3 is an odd number, and you can't divide 10 into 3 equal groups.

**Aaron:** 33 plus 33 plus 33 equals 99, so how can you make 100?

**Ricardo:** I think if you take all those 30's and make 'em into 20's, it would work.

**Jennifer:** If you start with 5's, it might do it.

By the end of this discussion, some students believed that there is no solution with whole cents, while others were still convinced that sharing in the right way might make it possible to divide $1.00 in three ways evenly.

Students then worked with interlocking cubes to model this problem. Those who were pretty sure that there was no way to divide $1.00 evenly among three people enjoyed proving their solution with the cubes, while students who were unconvinced by the group discussion needed to work with the cubes themselves.

Tamara's conjecture, "You can't divide an even number by an odd number," is an idea that often comes up in this discussion. While Tamara is right that you can't divide 10 into three equal groups, her generalization does not hold up—for example, you can divide 10 by five. If this idea comes up in your class, you might want to list the conjecture somewhere, ask students to think about it, and discuss it further when students work on division problems later in the unit: "Can we ever divide an even number by an odd number? an odd number by an even number?"

# INVESTIGATION 2

# Using Landmarks to Solve Problems

## What Happens

**Sessions 1, 2, and 3: Moving Beyond 100** After reviewing what they have found out about the factors of 100, students use this knowledge to explore larger numbers. Through a series of activities set up as choices, they explore the multiples of 100 with different tools, including money, the 300 chart, and the calculator. They solve problems involving multiplication and division by 20, 25, and other familiar factors, in the context of money. They examine the patterns in counting by 20's to 1000, and they begin to develop strategies for division using these patterns.

**Session 4: Solving Problems with Money** After a whole-class discussion of one multiplication problem, students work in pairs on multiplication and division problems that involve money. They work carefully on one, two, or three problems, writing and drawing to explain how they found their solutions.

**Sessions 5 and 6: Real-World Multiplying and Dividing** These sessions begin with an introduction to standard division notation. Students then apply their understanding of landmarks to multiplication and division problems in real-world contexts. Through a series of activities set up as choices, they solve given problems, find groups of objects in the classroom that are close to multiples of 100 (e.g., 30 students with 10 fingers each), and create and solve their own problems. They use multiplication and division notation to record their work, and learn to recognize a variety of symbols used to write down multiplication and division. They also become familiar with how to do multiplication and division on the calculator.

## Mathematical Emphasis

- Using knowledge about factors of 100 to understand the structure of multiples of 100 (if there are four 25's in 100, there are twelve 25's in 300)

- Developing strategies to solve problems in multiplication and division situations by using knowledge of factors and multiples

- Estimating real quantities that are close to 200, 300, and 400

- Reading and using standard multiplication and division notation to record problems and answers

How many quarters
are there in $4.25?

○  There are 8 quarters
   In two dollars.

   ○ ○ ○ ○
   ○ ○ ○ ○
   8 + 8 = 16 which
○  equals four dollars.

   There is 1 quarter
   in 25 cents.

○  1 + 16 = 17 and THAT
   is the answer

## What to Plan Ahead of Time

### Materials

- All materials used previously in the unit remain available: cubes, coins, and 100 charts. Make sure students know that they may use any of these materials that they would find helpful.
- Calculators: at least 1 per pair (all sessions)
- "Pictures of 100" that students made in Investigation 1 (Sessions 1–3)
- Overhead projector (Sessions 1–3)

### Other Preparation

- Duplicate student sheets and teaching resources (located at the end of this unit) in the following quantities. If you have Student Activity Booklets, copy only the items marked with an asterisk, including any transparencies needed.

#### For all sessions

300 chart (p. 91): 5–7 per student (2 for homework), and 1 overhead transparency*

#### For Sessions 1–3

Student Sheet 11, Exploring Multiples of 100 (p. 79): 1–2 per student

Student Sheet 12, Calculator Skip Counting (p. 80): 1 per student

Money Problems* (p. 85): 5 copies, cut apart, and sets placed in boxes or envelopes

Student Sheet 13, How Many 4's Are in 600? (p. 81): 1 per student (homework)

#### For Session 4

More Money Problems* (p. 86): 5 copies, cut apart, and sets placed in boxes or envelopes

#### For Sessions 5–6

More Money Problems* (p. 86): 5 copies, cut apart, and sets placed in boxes or envelopes

Student Sheet 14, Multiplying Things in Class (p. 82): 1 per student, or 1 to post for student reference

Division Problems* (p. 87): 5 copies, cut apart, and sets placed in boxes or envelopes

Student Sheet 15, Multiplying Things at Home (p. 83): 1 per student (homework)

Student Sheet 16, Making Up Landmark Problems (p. 84): 1 per student (homework)

# Moving Beyond 100

## Materials

- Students' pictures of 100
- Cubes, coins, and 100 charts
- Money Problems, in sets
- Student Sheet 11 (1–2 per student)
- Student Sheet 12 (1 per student)
- Transparency of 300 chart
- Overhead projector
- 300 chart (4–5 per student, 2 for homework)
- Calculators
- Student Sheet 13 (1 per student, homework)

## What Happens

After reviewing what they have found out about the factors of 100, students use this knowledge to explore larger numbers. Through a series of activities set up as choices, they explore the multiples of 100 with different tools, including money, the 300 chart, and the calculator. They solve problems involving multiplication and division by 20, 25, and other familiar factors, in the context of money. They examine the patterns in counting by 20's to 1000, and they begin to develop strategies for division using these patterns. Their work focuses on:

- using what they know about 100 to think about multiples of 100
- proving how many groupings of a particular factor make 100

 **Ten-Minute Math: Calendar Math** Once or twice during the next few days, do Calendar Math. Remember, Ten-Minute Math activities are designed to be done outside of the math hour.

Ask students to give number combinations that equal the number of the day's date, using multiplication as part of their answer.

List student responses. For example, if the date is January 21, solutions could include $2 \times 10 + 1$, $3 \times 10 - 9$, or $5 \times 5 - 4$.

Choose a "favorite expression" for the day, perhaps the most unusual, or one that uses a new idea. Use this favorite to write the date on the board, for example: January $0 \times 21 + 21$.

For variations, see p. 63.

## Activity

### Hundreds from Home

We've been working with the number 100 recently, and we're going to be working with even higher numbers soon. Let's start today by looking at the pictures of 100 you made earlier.

Ask several students to show the arrangements of 100 objects or pictures they made at school and at home. Have them describe how they arranged the items to be able to "prove" that there are exactly 100 objects without counting them one by one.

Spend a bit of time adding by 100's, using the student displays as concrete objects to add. See the **Dialogue Box**, Hundreds from Home (p. 39), for an example. Ask questions such as:

**How many 20's are there altogether if we put Jamal's and Annie's together?**

## Choice Time: Exploring Multiples of 100

**Four Choices**   During Sessions 1–3 of this investigation, students may choose from four activities going on in the classroom simultaneously. This format allows students to explore the same idea at different paces. Some might try two activities during each hour mathematics class. They may repeat an activity, using different numbers or problems. They need not try all four choices.

**How to Set Up the Choices**   If you set up your choices at stations, show students what they will find at each station. Otherwise, tell students where they can find the materials they need.

> Choice 1: Factors of Numbers Greater Than 100—copies of Student Sheet 11, cubes, and coins
>
> Choice 2: Money Problems—sets of Money Problems in boxes or envelopes
>
> Choice 3: Counting on the 300 Chart—copies of the 300 chart
>
> Choice 4: Calculator Skip Counting—calculators and copies of Student Sheet 12

Introduce the four activities. Choice 3, which involves the 300 chart, and Choice 4, which entails calculator skip counting, both need to be introduced with some whole-class work; see the following activity descriptions for specifics. (You may want to delay the introduction of Choice 4 until the beginning of Session 2.) Students then begin working on the activities of their choice.

You might list the four choices on the board, as a reminder for the students.

### Choice 1: Factors of Numbers Greater Than 100

Students take a copy of Student Sheet 11, Exploring Multiples of 100, along with cubes and coins for making groupings. Working alone or in pairs, students choose 4, 5, 10, 20, or 25 as a factor. They find out how many of their chosen number are in 100, how many are in 200, and how many are in 300.

Even if working in pairs, students fill out their own copies of Student Sheet 11. When completing the statement "I know there are _____ in 300 because ...," their sentence should have enough information to convince someone else how many of their factor are in 300. For examples of student work, see the illustration on p. 35.

There is space on Student Sheet 11 for students to continue past 300 if they want. These students can complete the last section of Student Sheet 11, telling about any patterns they see. Some students may want to take a second student sheet and work on another factor after they have completed their work with the first.

### Choice 2: Money Problems

Students choose problems from the sets of Money Problems. They do not need to do all the problems; you can adjust your expectations for individual students. Students are to write about each problem on a separate piece of paper or in their mathematics notebook, telling how they solved it.

### Choice 3: Counting on the 300 Chart

Use the overhead transparency to introduce the 300 chart. Ask students to suggest one way to skip count on the chart that will land them exactly on 300. Ask how they know that number will work, and what they think the skip counting pattern for that number will look like on the chart. Mark the count for this number on the transparency as an example.

In this activity, students use copies of the 300 chart and find numbers that will land them exactly on 300 if they count by that number—in other words, they find factors of 300. They mark a 300 chart for each factor they find.

You may want to reread the **Teacher Note** in Investigation 1, Students' Problems with Skip Counting (p. 14); students may encounter similar problems in using the 300 chart.

### Choice 4: Calculator Skip Counting

Introduce calculator counting to the whole class. Ask students to contribute ideas about different ways to make the calculator skip count by 10's up to 300. (For basic instructions on making a calculator skip count, see the Ten-Minute Math activity on p. 17.) Give students some time to practice calculator skip counting by different numbers.

Make available copies of Student Sheet 12, Calculator Skip Counting. Students use the calculator to count by a number that will get them exactly to 300. Their task has three parts: to find a way to make the calculator count by this number, to keep track of how many of that number it takes to make 300, and to record their results on Student Sheet 12.

### While Students Are Working on the Choices

- At some point during these three sessions, pause for the class discussion activity, Finding Patterns in the 20's (p. 35). Afterward, students may continue with the activities.
- During Session 3, have each student spend some time on the Assessment activity, How Many in 500? (p. 36).

Hold this discussion sometime during Sessions 1–3, after students have had time to try several of the activity choices. Choose a factor of 100 with which your students are quite comfortable. Using this factor, begin a chart, like this one for the factor 20.

| Number | How many 20's? |
|--------|----------------|
| 100    | 5              |
| 200    | 10             |
| 300    | 15             |

Ask students to look for patterns.

**What do you notice about the number of 20's as you look down the chart? How many 20's would there be in 400? How do you know?** [*Add this result to the chart.*] **Can you predict for higher numbers?** [*Add 500 and 600 to the chart.*]

**How many 20's are in 900? Why do you think so?** [*Pause on this question until many students have given their own explanations.*]

**How many 20's are in an "in-between" number, like 640? Does 20 work for 640? How do you know?**

This is an important discussion. Plan to spend about 20 minutes on it. See the **Dialogue Box**, How Many 20's in 280? (p. 40), for an example of a third grade discussion along these lines.

---

I know there are ___75 4's___ in 300 because  since there are 25 4's in 100, so I just counted it like quarters. And it came out 75.

I know there are ___30 10's___ in 300 because 10 plus 10 is 20, 20 plus 10 is 30.

I know there are ___fifteen 20's___ in 300 because there are five in 100 so you add five three times, but we still used cubes to make sure.

I know there are ___60 5's___ in 300 because I put a cube on the 300 chart after every 5, and then I counted them, and it was 20 for 100, 40 for 200, 60 for 300. But I also know another way because 20 and 20 and 20 is 60.

I know there are ___12 25's___ in 300 because I know that in 3 dollars there are 12 quarters, so there must be 12 25's in 300.

---

*Typical student explanations of their reasoning on Student Sheet 11.*

**Assessment**

**How Many in 500?**

Toward the end of Sessions 1–3, ask students to work on another problem:

**How many 5's (or 20's, or 25's) are in 500?**

Since they don't have enough cubes or coins or counting chart squares to actually count to 500, they will need to use their knowledge about 100, 200, and 300 to visualize what happens as they go on to 400, then to 500.

You might assign different factors to different pairs of students; most should be working with 5, 20, or 25. Students who do not seem ready for this assignment could find the number of 10's in 500. Students needing a more difficult challenge can find how many 4's are in 500.

When pairs of students are finished, ask them to write down their solution or to explain their solution to you. Tell them you will be looking at every-one's solution. Because some students may not be adept at conveying their thinking in writing, it is important that you spend enough time with every individual to assess their strengths and needs.

Use these questions to guide your observation and questioning, both during this assessment and during their other work in Sessions 1–3:

■ Are students able to skip count confidently, knowing where to start, and making use of visual or numerical patterns to ensure accuracy?

■ Do they make appropriate use of tools (coins, cubes, counting charts) to solve these problems?

■ Can they build on partial solutions (such as their knowledge of the num-ber of 5's in 100) to develop solutions for higher numbers (the number of 5's in 500)?

■ Do they notice and correct their own mistakes? Do they double-check their counting strategies?

Refer to the **Teacher Note**, Assessment: How Many in 500? (p. 37), for commentary on some sample student responses.

## Sessions 1, 2, and 3 Follow-Up

**Homework**

**Money Problems**   After Session 1, students take home Money Problems they have not yet done (from the class sets), or similar problems that you create.

**Skip Counting**   After Session 2, students should take their 300 chart home to look for other numbers that land exactly on 300 when they skip count by that number.

**How Many 4's Are in 600?** After Session 3, give students a copy of Student Sheet 13, How Many 4's Are in 600? and two 300 charts.

Students write a sentence to show how they figured out this problem, and draw a picture to explain their answer to another student. You can adjust the difficulty level of this problem for various students. Counting by 4's is harder than 5's or 25's, and 10's is easier than any of these.

## Assessment: How Many in 500?

The assessment question, "How many 5's (or 20's, or 25's) are in 500?" is a difficult problem for many students, and you may not see many perfect solutions. However, students' approaches to the problem will give you a great deal of information about their familiarity with the factors of 100, their facility with skip counting, and their understanding of the relationship between 100 and its multiples. The following examples of student responses demonstrate the process.

### Maria: How Many 20's in 500?

To explain her answer, 25, Maria writes:

> There are 5 20's in 100, so do it 5 times and it's 25.

**Can you explain it a little more?**

**Maria:** It's easy. You know there are five 20's in 100.

**How do you know?**

**Maria:** Because [*showing one finger for each number she counts*], 20, 40, 60, 80, 100. See, five twenties. Then, it's just the same for 200, so it's five more, then five more for 300, until you get to 500. That's 5, 10, 15, 20, 25 [*again, she shows one finger for each count*].

The teacher wants to see whether Maria can use this method with other numbers.

**What would happen for 5's? How many 5's in 500? How could you do it?**

**Maria:** Well, you'd do the same thing. Find out how many in 100, then just count that 5 times.

**Do you know how many are in 100?**

**Maria** [*thinks for a minute*]: Um, I'm not sure, but I could figure it out if I counted by 5's.

Maria is comfortable with skip counting by 20's and with using the method of skip counting to find out how many of a particular number are in some multiple of that number. She is able to keep track of her work; she "double counts" easily—that is, she counts by 20's while also keeping track of the number of 20's she is counting. She has a clear grasp of how 100 is related to its multiples, and she can use this information to solve problems. However, she does not use her knowledge that five 20's make 100 to figure out how many 5's are in 100.

### Aaron and Ricardo: How Many 4's in 500?

The teacher gives Aaron and Ricardo this slightly more difficult problem, having already seen them solve quite easily the problem that Maria worked on. The two students decide that there are 204 fours in 500. They use the 100 chart and explain their solution this way:

> See, we counted by 4's and got 28, and then we doubled it and got 56, and then we got 102, and then 204!

*Continued on next page*

From their work and some further questioning, the teacher notices that they have good strategies for adding numbers, and that they understand how to skip count in order to figure out how many of a number are in one of its multiples. However, they have difficulty keeping track of their count, and when they make an error in counting, they don't catch it.

This is a pattern the teacher has noticed in Aaron and Ricardo's work—some elegant strategies that are undermined by small errors and a lack of checking. The teacher makes a mental note to emphasize the use of more than one strategy to help Aaron and Ricardo double-check their work. Then the teacher asks them how they know they have the correct number of 4's for 100. When they say, "We counted," the teacher asks if they could use the pattern on their 100 chart to see if they counted correctly.

### Tamara and Jamal: How Many 20's in 500?

This pair of students at first seem baffled by the problem. They have a hard time getting started, and when the teacher asks them what they are thinking, Tamara says, "We can't do it. We don't have enough blocks." The teacher asks them to figure out how many 20's are in 100, and the two set to work building.

First they count out 100 cubes, then begin taking groups of 20, counting by 1's. Then the teacher comes around to their workspace again, asking "Is there any way you can make your 20's by counting by a bigger number?" Jamal replies, "By 10's!"

Since many of the interlocking cubes are already put together in groups of 10, the two students quickly assemble groups of 20's. They count the groups of 20's by 10's until they get to 100, and write down on their paper:

There are 5 20's in 100. This is how we know. We have 20 + 20 + 20 + 20 + 20 and that makes 100.

The teacher asks them to find the number of 20's in 200, and they use the same method, making 20's until they reach 200, counting by 10's to make sure they have exactly 200, then counting the number of groups of 20.

Tamara and Jamal need a lot more experience with counting by 10's and 20's and answering questions about 100, 200, and 300. They are comfortable building with cubes, but need to do more counting on the 300 chart as they construct their cube groups, so that they begin to associate the numerical patterns with the groups they are counting. They need to become comfortable counting by numbers larger than 10 and to see how the things they do in the first hundred can help them with what happens in the next hundred.

# D I A L O G U E   B O X

## Hundreds from Home

In this discussion during the first session of Investigation 2 (p. 32), the class is talking about the pictures of 100 the students made at home.

**What did Ryan do here that's extra-special?** [*Ryan has glued 100 metal soda can pop-tops onto a piece of cardboard.*]

**Chantelle:** He recycled.

**That's what I was thinking too. How did you arrange them, Ryan?**

**Ryan:** I did them by 10's. We have tons of them at home.

**Look at this display of popcorn kernels, 10 by 10. What could we call that?**

**Seung:** Rows and columns.

**Dominic:** It's like a 100 chart.

**Yes, it's very organized. Dominic, tell me about your display.**

**Dominic:** I cut out little squares and I put them by 5's in an X, the way the five is on dice.

**Laurie Jo did her stamps in 5's also. How did you arrange them?**

**Laurie Jo:** I have 20 rows of 5's, and that makes 100.

**If we put Dominic's and Laurie Jo's charts together, what we would have?**

**Ricardo:** 200!

**Jamal:** There would be twenty 5's.

**Su-Mei:** Ten 5's make 100, so twenty 5's for 200.

**If we put the popcorn with the pop-tops, what would we get?**

**Jeremy:** 200.

**And if we put these macaroni pieces with them?**

**Yvonne:** 300.

[*The teacher points to two student displays made from 100 bits of cut-up paper.*]

**What if we put this big batch of paper with this other big batch?**

**Annie:** 200.

**And what if we put everyone's 100's together?**

**Elena:** Twenty-one hundred?

**Why do you think that?**

**Elena:** Because there are 22 people, and everybody brought something but Mark. And everybody had one hundred.

**What does anyone else think?**

**Ly Dinh:** I think you could count by 100. It would be ten, then ten more, and one extra, so it would be twenty-one hundred.

**Is there any other name for twenty-one hundred?**

**Chantelle:** Twenty hundred and one?

**Mark:** Two thousand and one.

**Michael:** Two thousand and one hundred.

At this point, this teacher chose to list all these ideas about naming this number on a piece of chart paper to return to later during the last investigation in this unit.

## How Many 20's in 280?

A third grade class has created a class chart to show how many 20's are in multiples of 100, as explained in the activity Finding Patterns in the 20's (p. 35). Here they are discussing patterns in their chart.

| Number | How many 20's? |
|--------|----------------|
| 100 | 5 |
| 200 | 10 |
| 300 | 15 |
| 400 | 20 |
| 500 | 25 |
| 600 | 30 |
| 700 | 35 |
| 800 | 40 |
| 900 | 45 |
| 1000 | 50 |

**What if we listed an in-between number, like 640? First we have to decide, is 640 a multiple of 20?**

**Students** [*variously*]: Yes … No … Yes …

**Michael:** Because 20 + 20 is 40, and 600 has 30 twenties in it.

**Rashad:** What did he say?

**Michael:** Because we already know that every 100 has 5 twenties, so that's up to 600. Then you just count by 20's—20, 40—and you get to 640.

**What about 710?**

**Jennifer:** I think it's the same thing. You can do 20's to 700, and then … oh, I don't know.

**Su-Mei:** Just see if 10 is a multiple of 20. [*She checks on her miniature 100 chart on which she counted by 20's.*] Because 10 is BEFORE 20, so there's no way it could be a multiple.

**How about 360? Is it a multiple of 20?**

**Ryan:** Yes, because if you count by 20 you land on 60.

Sensing that some of the students are not following this discussion as well as the more vocal ones, the teacher asks everyone to think about how many 20's are in 280. The students are encouraged to talk among themselves. After a few minutes, the teacher asks for their solutions.

**Amanda:** Four 20's represents 80, and there are ten 20s in 200. Then you add 10 + 4, and it's 14.

**Tyrell:** I used the chart. So there were ten 20's in 200. Then I counted 20, 40, 60, 80.

**Midori:** I drew some boxes to represent 20's. I just guessed at how many there were at first. Then I skip counted each box by 20's. Then when I got to 280, I stopped and counted the boxes.

**Jennifer:** Seung helped me. First I told her that four 20's make 80. Then she asked me how many 20's were in 100 and I told her 5. Then she asked how many in 200 and I knew it was 10. So I added it and got 14.

**Saloni:** I counted on my fingers how many 20's in 100, and there were 5, and I knew there were four 20's in 80, and I added them.

**Liliana:** I drew five 20's and it went 20, 20, 20, 20, 20. And every time I got five, I put a circle and wrote how much it was worth, and I got 14.

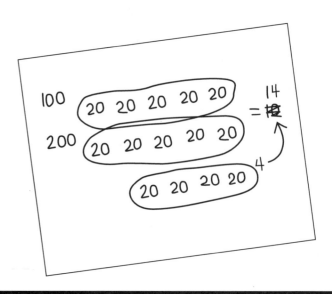

# Solving Problems with Money

## What Happens

After a whole-class discussion of one multiplication problem, students work in pairs on multiplication and division problems that involve money. They work carefully on one, two, or three problems, writing and drawing to explain how they found their solutions. Their work focuses on:

- using their knowledge of factors to solve problems in multiplication and division situations
- recording their solutions in a way that explains their thinking to others

### Materials

- Cubes, coins
- 300 charts
- More Money Problems, in sets
- Calculators

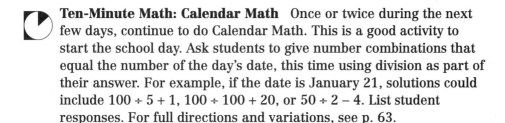 **Ten-Minute Math: Calendar Math** Once or twice during the next few days, continue to do Calendar Math. This is a good activity to start the school day. Ask students to give number combinations that equal the number of the day's date, this time using division as part of their answer. For example, if the date is January 21, solutions could include $100 \div 5 + 1$, $100 \div 100 + 20$, or $50 \div 2 - 4$. List student responses. For full directions and variations, see p. 63.

### Activity

Pose the following problem :

> At a certain school, lunch costs $1.50. If six people buy lunch one day, how much money will the teacher have after collecting their money?

Or, use other familiar real-world situations: at a museum, admission is $1.50; bus fare is $1.50; and so forth.

**Note:** Always feel free to substitute or add problems from real-world contexts familiar to your students. Many of the problems in this unit were invented by teachers, based on situations familiar to their students.

Give students a few minutes to talk among themselves or to jot down or sketch a solution. Then ask them to share their strategies. Encourage students to share the way they started, even if they didn't fully solve the problem. See the **Dialogue Box**, Six People with $1.50 Each (p. 43).

## Class Discussion: Six People Paying $1.50 Each

## Problems Using Money

Make available the sets of More Money Problems, in boxes or envelopes. Students choose one, two, or three of these problems or similar problems you have created. If you create your own, include a mixture of multiplication and division situations. You may want to modify the given problems, using larger or smaller numbers, to meet the needs of individual students.

Students can work in pairs to solve the problems, but each student records the solution for each problem. Some pairs of students will require a whole session to do one problem. Encourage them to use cubes, coins, or 300 charts.

The point of doing only a few problems is to give students time to work out their solutions, double-check them, and write or draw to describe how they got each solution. Students are expected to find their own ways of recording their work on a plain sheet of paper. As you interact with students, make sure you can understand their sketch or explanation; if you can't, ask them to revise it to make it clearer.

## Session 4 Follow-Up

 **Homework**

**More Money Problems**   Give students an additional problem from the sets of More Money Problems to do as homework.

# D I A L O G U E ☐ B O X

## Six People with $1.50 Each

This discussion occurred after students had worked on the money problem posed in the activity on p. 41.

**How did you figure out how much $1.50 lunch money, collected from six people, is altogether?**

**Liliana:** I put two lunches together, then I added—$3.00, $6.00, $9.00.

**Who did it a different way?**

**Cesar:** I drew 6 lines.

**What did they represent?**

**Cesar:** One dollar, so that was 6 dollars. And then I put six 50's and kept adding two 50's to make more dollars. I got 3 dollars and added them to the 6 dollars I had.

**Amanda:** I had $6.00, then I went [*holding up a finger for each 50 cents*] $6.50, $7.00, $7.50, $8.00, $8.50, $9.00.

**Dylan:** I had 6 dollars, one dollar for each $1.50. Then I know that if I had 4 more dollars, it would be 10. But 6 more 50 cents's only makes 3 dollars, so that's one less than 4, so it's 9.

**Khanh:** On my paper I put 6 lines for 50's and then circled groups of dollars and added them on to 6.

**Jennifer:** I drew 50 cent pieces and said two 50's is $1.00 and so on, and I came up with $9.00.

# Real-World Multiplying and Dividing

## Materials

- More Money Problems, in sets
- Coins, cubes, 300 charts
- Calculators
- Student Sheet 14 (1 per student, or 1 posted)
- Division Problems, in sets
- Student Sheet 15 (1 per student, homework)
- Student Sheet 16 (1 per student, homework)

## What Happens

These sessions begin with an introduction to standard division notation. Students then apply their understanding of landmarks to multiplication and division problems in real-world contexts. Through a series of activities set up as choices, they solve given problems, find groups of objects in the classroom that are close to multiples of 100 (e.g., 30 students with 10 fingers each), and create and solve their own problems. They use multiplication and division notation to record their work, and learn to recognize a variety of symbols used to write down multiplication and division. They also become familiar with how to do multiplication and division on the calculator. Their work focuses on:

- developing their own strategies to solve problems, using their knowledge of factors and multiples
- using standard multiplication and division notation to record work
- using the calculator to do multiplication and division problems

 **Ten-Minute Math: Calendar Math**  Continue using Calendar Math in short sessions outside the mathematics hour. Ask for combinations that involve division, as described in Session 4. Consider having students work individually or in pairs with calculators as they look for new ways to make the day's date. For full directions and variations, see p. 63.

## Activity

### Using Standard Notation for Division

Pose a division problem in words, based on a real situation that is familiar to your students. Keep the numbers small and familiar, and don't tell your students that this is division. Here are some examples you can use or modify:

> Suppose I'm building toy cars. I bought a box of 24 little wheels to use on the cars. How many cars could I build with the 24 wheels?

> I'm going to share a little snack-size box of raisins among three people—myself and two friends. I counted the raisins in the box and found out there are exactly 30 raisins. How many can each person have?

> Suppose I wanted to divide this class into two teams. How many would be on each team?

Start with one problem. Ask students to solve it and briefly explain their solutions. You may want to encourage the use of cubes for a concrete representation of the problem.

Then ask if the students know some ways to write the problem down, without its solution, using numbers and symbols. Have everyone try to write down one way. Reread the **Teacher Note**, What About Notation? (p. 21), for more about the relationship between student strategies and standard notation.

Write the problem on the board in words, and then record student suggestions for how to use the symbols. You may end up with many different ways, some of which are correct and some of which are not.

How many cars can I make with 24 wheels ?

$$4 \div 24$$

$$\boxed{24 \div 4}$$

$$\widetilde{4 ) 24}$$

$$24 ) 4$$

Acknowledge that writing down division is very confusing. Show the two correct ways of writing the problem, and then show how to record the solution each way.

$$24 \div 4 = 6 \qquad 4 ) \overline{24}^{\ 6}$$

Point out that these two different forms can be hard to remember because they are read in different directions, and the numbers are in different places. Ask students to describe the differences between the two notations, and ask them if they have any ideas about how they can remember how to write each one. For more about division notation, see the **Teacher Note**, Talking and Writing About Division (p. 50).

Try another problem or two. Have students solve the problem and then write down the problem and solution using numbers and division symbols. They will have the chance to practice using standard notation for recording during the next few activities.

**Note:** Some students may also use the fractional forms of recording division:

$$^{24}/_4 \qquad \frac{24}{4}$$

Acknowledge those as correct, but focus on the other two forms for now.

# Choice Time: Working with Landmarks

**Four Choices**   During Sessions 5–6 of this investigation, students may choose from four activities going on in the classroom simultaneously. This flexible format allows students to explore the same idea at different paces. Students may try all or just some of the activities, and they may repeat those that work best for them, using different numbers or problems.

**How to Set Up the Choices**   After introducing standard notation for division with the preceding activity, introduce the four activities students may choose from for the remainder of Session 5 and Session 6. If you set up your choices at stations, show students what they will find at each station. Otherwise, tell students where to get what they need for each choice.

> Choice 1: More Money Problems—sets of More Money Problems in boxes or envelopes
>
> Choice 2: Multiplying Things in Class—copies of Student Sheet 14 (or 1 copy, posted)
>
> Choice 3: Division Problems—sets of Division Problems in boxes or envelopes
>
> Choice 4: Making Up Landmark Problems—paper and pencils

In all four activities, encourage students to use their knowledge of factors and multiples to solve the problems.

You might list the four choices on the board, as a reminder for the students.

## Choice 1: More Money Problems

Students continue to work on problems they have not yet solved from the sets of More Money Problems. They may use cubes or coins if they want.

## Choice 2: Multiplying Things in Class

Instructions for this activity are given on Student Sheet 14, Multiplying Things in Class. Make copies available, or simply read and post the sheet. The goal is to find and count large numbers of things in the classroom that come in groups, such as fingers, boxes of crayons, chair legs, or floor tiles. Students can record individually what they find. Or, you can make a class list of what they find and how many there are of each thing.

If you have packages with the same number of items in each box (for example, paper clips), you might want to put them out for students to find. Students could also add things from another room to their count; for example, the number of chair legs in two classrooms.  For different approaches students may take, see the **Teacher Note**, Using Multiples to Count (p. 49).

**Note:** In this activity, some students may be ready to use their knowledge of landmark numbers to reason about numbers that are close to these landmarks. For example, consider the problem, "How many playing cards are there in three decks?" The number 52 (cards per deck) is close to the landmark number 50. So, students might combine their knowledge of counting by 50's with their knowledge of counting by 2's to solve the problem $52 \times 3$. For an example of one student's experience with near-landmark numbers, see the **Dialogue Box**, How Many Teeth in Our Class? (p. 51).

### Choice 3: Division Problems

Students work on one or two of the problems from the Division Problems sets and record with words or pictures, on another sheet of paper, how they solved the problem. They also write the problem and solution using standard division notation.

You can make up additional problems to provide an appropriate level of challenge for individual students. Some students may need more practice with problems involving familiar factors (20, 25, 5, and so forth), while others can begin to branch out to problems involving less familiar factors (for example, 15 or 30).

### Choice 4: Making Up Landmark Problems

Students make up their own problems, like the ones they've been working on, and illustrate them. You might establish three rules:

1. Each problem must use important landmark numbers, like 20, 25, 75, 100, and so forth.
2. Each problem must be a real, or at least realistic, situation.
3. You must solve every problem you write. Use a separate piece of paper for the solution.

Students enjoy having their problems (without the solutions) put together into a class book. This book can be sent home and can be used for additional practice.

**While Students Are Working on the Choices**  At some point during these two sessions, interrupt student work for the following:

■ Share and discuss some of the multiplication problems students have created from things in the classroom (Choice 2). What did they find? How did they count the items?
■ Do the activity Multiplying and Dividing on the Calculator (p. 48) with the whole class or small groups.

## Multiplying and Dividing on the Calculator

Sometime during these two sessions, encourage students to try some simple multiplication and division problems on the calculator. If you have enough calculators, you can do this with the whole class. Students can work with calculators individually or in pairs, taking turns pressing the keys. Otherwise, you can do this with smaller groups as students are working on other choices.

Pose a real-world problem, using small and familiar numbers (such as the toy car problem suggested earlier). Ask students to solve this on their calculator.

**What keys did you need to press to solve the problem? Are the symbols on the calculator similar to the standard notation for multiplication and division? Is the order in which you press calculator keys similar to the way you write down a problem?**

If you have a variety of calculators in the classroom, students can compare them to see if they work in the same way.

## Sessions 5 and 6 Follow-Up

  **Homework**

**Multiplying Things at Home**   Multiplying Things in Class (Student Sheet 14) has been adapted for homework on Student Sheet 15, Multiplying Things at Home. Ask students to look for large numbers of things at home, record what they find, and describe how they figured the total.

**Making Up Landmark Problems**   Choice 4, Making Up Landmark Problems, can also be continued as homework on Student Sheet 16. Students can ask their family members or friends to solve the problems they invent.

I have 300 peanuts in a bag and 25 kids in my class. How can I split the peanuts so we all get the same?

**Things That Number in the Low Hundreds** Once students have had some experience with things that number close to 200, 300, or 400, they might be able to predict other things found in these quantities around the school, the neighborhood, or at home. Make a list of students' ideas. Give them some time (either during class or as homework) to count the things in order to verify their ideas. If you started a class list of the quantities students found in the Choice 2 activity, you can now add the other things they find.

**Things That Number in the High Hundreds** Ask students to think of (or find) things that number close to even greater multiples of 100. That is, what can they think of that might be counted as 500? 700? 1000? This search is also a good opportunity to involve students' families: Find things at home that total close to 500. This work will connect to activities in Investigation 3 of this unit.

**Exploring Different Calculators** Ask students to explore any calculators they may find at home. Are the symbols the same as those on the ones they use in school? Can they do multiplication and division problems the same way? Students might draw a picture of any calculator that looks different to them and share these differences with the class.

## *Using Multiples to Count*

**Teacher Note**

In these sessions, students find examples of things that give them a chance to visualize 100, 200, and 300, and that let them explore ways they can figure out multiplication situations such as "30 chairs, with 4 legs each." There are several good ways to count the legs of 30 chairs.

■ Some students may note that there are 25 fours in 100 and then 5 more fours, making 120.

■ Others might use the 100 or 300 charts to skip count by four 30 times.

■ Some might count by four 10 times, getting to 40, then add 40 three times on the calculator.

Encourage students who are only counting by 1's to begin to take some shortcuts, based on what they know about skip counting from previous sessions.

Various division symbols are used as standard notation in our society:

$$24 \div 4 \qquad 4\overline{)24} \qquad {}^{24}/_4 \qquad \frac{24}{4}$$

There are also different ways to "read" or speak of these notations:

1. Four goes into (or, as students say, "guz-inta") 24

2. 24 divided by 4

3. How many 4's are in 24?

4. When 24 is shared among 4 people, how many does each person get?

So many symbols and so many different ways of reading them can be very confusing to young students. Especially confusing is the fact that the numbers and symbols appear in different positions, depending on the notation you are using.

In this unit, we ask students to recognize only the first two forms of notation. The first two ways of "reading" these notations correspond to the ways the symbols are written, and may seem simpler to you. However, "goes into" does not convey any mathematical meaning, and "divided by" may not be easily understood by students who don't yet understand what division is. The third and fourth ways of speaking carry more meaning for young students.

In this unit, we want to focus on the "grouping" interpretation of division: "How many 4's are in 24?" If students learn to read the two standard notations in this way, they will more likely attach meaning to the numbers and symbols.

For more about helping students connect their own good strategies with standard notation, see the **Teacher Note**, What About Notation? (p. 21).

# How Many Teeth in Our Class?

Mark, a third grader, is doing Choice 2: Multiplying Things in Class (p. 46). He has decided to figure out how many teeth there are in the classroom. Mark comes to his teacher with his paper showing a long column of 24's.

$$
\begin{array}{r}
24 \checkmark \\
24 \checkmark \\
96 \quad {}^+ 24 \checkmark \\
+ 24 \checkmark \\
+ 24 \checkmark \\
+ 24 \checkmark \\
96 + 24 \checkmark \\
+ 24 \checkmark \\
+ 24 \checkmark \quad \cancel{62} \\
96 \quad 24 \checkmark \quad 72 \\
+ 24 \checkmark \\
+ 24 \checkmark \\
+ 24 \checkmark \\
96 \quad 24 \checkmark \\
+ 24 \checkmark \\
24 \\
24 \\
24 \\
24
\end{array}
$$

**Mark:** I know that I have 24 teeth, and so I figured that everyone in my class is about my age, so they probably have about 24 teeth too. But I don't know how to add it up.

**Would it help if you thought of the 24's as 25?**

**Mark:** Oh! Well, four 25's is 100, so four 24's is, let's see, it's one less for each one … 96! [*Mark then groups his nineteen 24's into groups of four and writes 96 + 96 + 96 + 96 + 62.*]

**Where did the 62 come from?**

**Mark:** Well, I had three 24's left over, and that's 20, 40, 60—whoops!—that's 72, not 62.

**Is there a way you can think of those 96's so that they are easier to add—just like you thought of the 24's as 25's?**

**Mark:** They could be 100's—then I would just have to subtract four from each 100. That's 400 take away 16, so that's, let's see, take away 20 is 380, um … 384. [*Mark then adds 384 and 72 on the calculator, and says excitedly*] 456! That's a lot of teeth!

# Constructing a 1000 Chart

## What Happens

**Session 1: A 1000 Chart**  Students construct a poster-size chart of 1000 centimeter squares. Each pair constructs their 1000 out of a particular factor of 1000—either 20, 25, or 50—and labels each group up to 1000.

**Sessions 2 and 3: Finding Large Quantities** Students identify items that are found in large quantities in school or at home and relate those quantities to numbers on their 1000 chart. They then use the 1000 charts to calculate the "distance" (the difference) between numbers such as 550 and 950.

## Mathematical Emphasis

- Using factors of 100 to understand the structure of 1000 (how many 50's does it take to make 1000?)

- Estimating quantities up to 1000 (what can we find in the classroom that numbers about 500?)

- Using landmarks to calculate "distances" within 1000 (how far is it from 650 to 950?)

## What to Plan Ahead of Time

### Materials

- Large poster or chart paper: 1 sheet per pair  (Session 1)
- Scissors (Session 1)
- Glue or tape (Session 1)
- Crayons or markers (Session 1)
- Small counters or cubes, 1 cm or less: 1 per student (Sessions 2–3)
- String or ribbon (Sessions 2–3, optional)
- All materials used previously in the unit remain available: cubes, coins, 100 charts, 300 charts, calculators. Encourage students to use any of these materials that they would find helpful.

### Other Preparation

- Duplicate student sheets and teaching resources (located at the end of this unit) in the following quantities. If you have Student Activity Booklets, no copying is needed.

*For Session 1*

Student Sheet 17, More Than 300 (p. 88): 1 per student (homework)

One-centimeter graph paper (p. 89): 3–4 sheets per pair (or enough to give them 1000 squares, with extra for waste and experimentation)

- Construct a larger classroom 1000 chart with squares grouped by 20's or 25's and carefully sequenced. Use squares that are at least 1 inch or 2 centimeters on a side. Print the number for each block of squares (20, 40, 80 ..., or 25, 50, 75 ...) in the lower right square of the block. (Sessions 2–3, optional)

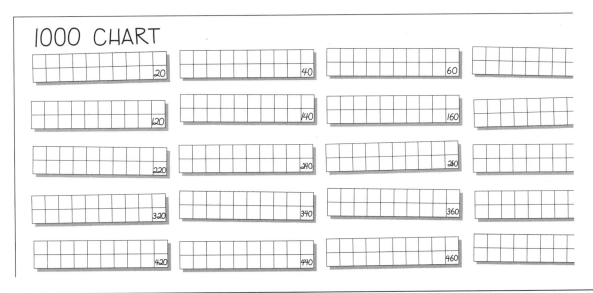

# A 1000 Chart

## Materials

- One-centimeter graph paper (3–4 sheets per pair)
- Large poster or chart paper (1 sheet per pair)
- Scissors
- Glue or tape
- Crayons or markers
- Cubes, coins, 100 charts, 300 charts, calculators
- Student Sheet 17 (1 per student, homework)

## What Happens

Students construct a poster-size chart of 1000 centimeter squares. Each pair constructs their 1000 out of a particular factor of 1000—either 20, 25, or 50—and labels each group up to 1000. Their work focuses on:

- Counting to 1000 by one of its factors
- Keeping track of how many groups are needed to make 1000
- Organizing and labeling groups clearly so that any number on the 1000 chart can be located easily

 **Ten-Minute Math: Counting Around the Class** Once or twice during the next few days, outside of math class, return to the short activity Counting Around the Class.

Try counting by 20's or 25's. Stop two or three times during the count and ask questions like this:

**We're at 450 now—how many students have we counted?**

After the count is complete, ask further questions:

**If we keep going, so that everyone has a second turn, how many students will have to have a second turn so that we reach 1000?**

For full directions and variations on this activity, see p. 61.

## Activity

### Making a Thousand

Give brief directions for this task: Each pair of students must cut out from the one-centimeter graph paper exactly 1000 squares and glue or tape them to the large sheet of poster or chart paper, to make a 1000 chart.

They can cut their squares out in groups of 20's, 25's, or 50's. That is, a pair who decides to make their chart by 25's will need to cut out 40 blocks of 25 one-centimeter squares, paste them to the large chart paper, and number each block—25, 50, 75, 100, 125, ... up to 1000.

If students ask, "How many squares are on the graph paper?" or "How many 20's will I need?" tell them that part of their job is to answer these questions for themselves. Let students figure out for themselves how to arrange their blocks of squares on the large paper, but remind them that they must arrange the squares so that someone looking at their chart could easily count the groups. They should be able to convince someone else that there are exactly 1000 squares on their chart when it is finished.

Third graders need about a one-hour class session to complete and label their chart. Some students may start labeling each group of 25 squares with 25, 25, 25 ... instead of 25, 50, 75.... Make sure they understand that they are to label each block of squares with the total so far, not just with the number they are counting by.

Circulate as students are working, asking questions that will push them to think mathematically about the process they are following. For example:

**How many squares do you have? How many more do you need? How many more 20's (25's, 50's) do you need?**

Ask pairs of students to review each others' 1000 charts to help correct misplaced numbers and to see if they can find numbers easily on each others' charts. If some students finish early, they can challenge each other to locate particular numbers on their charts (see Locating Numbers on the 1000 Chart, the first activity in Sessions 2 and 3).

## Session 1 Follow-Up

**More Than 300**   Challenge students to find a quantity of something at home that matches one (or more) of the numbers on their 1000 chart that's greater than 300. Students should record what they find on Student Sheet 17, More Than 300, for use in the activities in Sessions 2 and 3.

 **Homework**

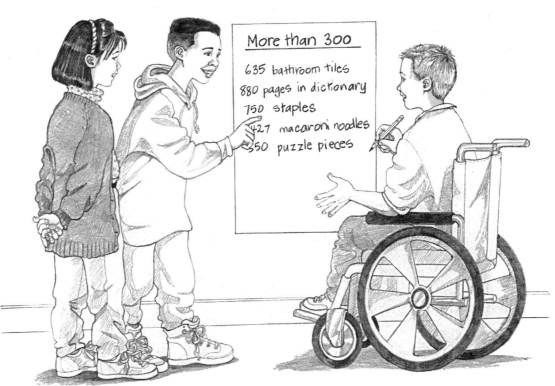

More than 300

635 bathroom tiles
880 pages in dictionary
750 staples
427 macaroni noodles
350 puzzle pieces

# Finding Large Quantities

## Materials

- Students' 1000 charts
- Small counters
- String or ribbon (optional)
- The large 1000 chart you made (optional)

## What Happens

Students identify items that are found in large quantities in school or at home and relate those quantities to numbers on their 1000 chart. They then use the 1000 charts to calculate the "distance" (the difference) between numbers such as 550 and 950. Their work focuses on:

- locating numbers between 1 and 1000
- estimating and counting real quantities up to 1000
- finding differences between two numbers by skip counting

 **Ten-Minute Math: Counting Around the Class** Continue to do Counting Around the Class at different times during the class day, as described in Session 1. Now that students are working with larger numbers, you might try counting by 50's or 100's. For other variations, see p. 61.

## Activity

## Locating Numbers on the 1000 Chart

Using the 1000 charts made in Session 1, ask students to locate particular numbers. Students should each have a small counter, piece of paper, or centimeter cube to place on their charts as you call a number.

For example, you might call "200." Students place their counters on 200 (or simply point to it) on their charts. You can circulate around the room quickly to check on students' accuracy and help those who are having difficulty.

At first, call out only multiples of 100. As students become more confident, call out multiples of the factors they have used—520, 375, 950, and so forth. Finally, call out any numbers up to 1000, such as 201, 335, 549, 857. Ask students to take turns picking numbers for the rest of the class to locate on their charts.

Repeat this activity several times during these two sessions. (One approach is to both start and end both sessions with this activity.) You may want to use the larger 1000 chart with the whole class or with small groups of students, especially if some of the students' charts aren't well organized.

# Finding Real Quantities for the 1000 Chart

Working in pairs, students try to find quantities of five to ten different items in the school that can be matched with numbers on the 1000 chart. For example, students might choose the number of books in the classroom, the number of students in the school, the number of tiles on the floor, the number of pages in a particular book, and so forth.

Before students start looking and counting, do at least one example with the class. Count the quantity, then discuss how you would find and label that quantity on the 1000 chart.

There are several different ways students might record each quantity that they find on their 1000 charts. They could write directly on the chart, or run a string or ribbon from the appropriate number on the chart to a description or picture of the thing they found. Some students like to set up a key, as on a map, using color or symbols to link the numbers and the items. For example, they might fill in particular number squares with different colors, explained in a key.

 532 students in the school

 360 boxes of pencils in the storeroom

 960 pages in the dictionary

You might establish the following guidelines:

1. Try to find something that matches six different numbers on your 1000 chart.
2. Make sure some of them are lower numbers, some are in the middle, and some are higher numbers.
3. Find at least one thing that is close to 1000.

Some students may want to use numbers they found for the activity in Investigation 2, Multiplying Things in Class, when they searched for grouped things in the classroom (fingers, crayons, chair legs). They may also include things they found at home (in the homework for Sessions 5–6).

As students are trying to find their own quantities in the classroom, encourage them to use efficient counting strategies—counting in groups, not by 1's. For example, suggest that they estimate by counting single handfuls or layers, or skip count if something is arranged in an array or equal groups.

If students want to investigate things outside the classroom (such as the number of bricks in the front hall, the number of wheels on cars in the school parking lot, the number of mail slots in the office, the number of students in the school), allow some time for student pairs to go count the particular quantities that interest them.

## Assessment

## Calculating with the 1000 Chart

After students have some confidence about finding quantities on the 1000 chart, have them practice finding "distances" on the chart.

**If I'm at 650, how far do I need to go to get to 1000?**

**How far is it from 325 to 1000? From 320 to 660?**

The purpose of these questions is not to evoke the standard subtraction algorithm, but rather to get students to count by units of 20 or 25 or 100 in going from one point to another, making use of landmarks.

Work on a few of these problems as a whole class, using the larger classroom 1000 chart. Then assign one or two problems to each pair of students. Every student chooses one problem to write up, showing the problem and explaining the solution in a way that would help someone else understand his or her thinking.

You can help students select problems at an appropriate level of difficulty. For example:

> Standard problem: How far is it from 350 to 800?
>
> Easier problem: How far is it from 300 to 450?
>
> Harder problem (with "in-between" numbers): How far is it from 480 to 740?

As you observe students' work in class and look over their solutions, you can get a sense of how well each student understands landmarks in the number system up to 1000. Use the following questions to guide your observations:

- Are students comfortable with reading and writing numbers in the hundreds?
- Can they skip count across hundreds?
- Can they use models to help them count (coins, cubes, counting charts)?
- Can they count comfortably by 25's and 50's, starting with a multiple of that number?
- Can they use skip counting patterns to help them solve problems?

Read the **Teacher Note**, Assessment: How Far from 650 to 1000? (p. 59) for more ideas on using this assessment.

As the unit ends, select one of the following options for creating a record of students' work on this unit.

**Choosing Student Work to Save**

- Students look back through their folders or notebooks and write about what they learned in this unit, what they remember most, what was hard or easy for them. You might have students do this work during their writing time.

- Students select one or two pieces of their work as their best work, and you also choose one or two pieces of their work, to be saved in a portfolio for the year. You might include students' written solutions to the assessment, How Many in 500? (Investigation 2, Session 3), and any other assessment tasks from this unit. Students can create a separate page with brief comments describing each piece of work.

- You may want to send a selection of work home for parents to see. Students write a cover letter, describing their work in this unit. This work should be returned if you are keeping a year-long portfolio of mathematics work for each student.

## Assessment: How Far from 650 to 1000?　Teacher Note

The assessment at the end of this unit (p. 58) checks students' ability to use and understand landmark numbers up to 1000. The following examples of third graders' solutions to this kind of problem show how you might judge their work.

### Latisha's Approach

Latisha easily figures out how far it is from 650 to 1000 by looking at the 1000 chart. She writes:

> I figured out that from 650 to 750 is 100, and if you keep going down to 950, it's 300. There were 50 left then, so I added it to 300 and that made 350.

Latisha seems comfortable with numbers in the hundreds and their relationships to each other. She can count comfortably by 100 from a number that is not a multiple of 100. We might like

to ask her more about how she knew that "from 650 to 750 is 100." Did she count? Did she "see" automatically that this distance is 100? Would she be able to figure out the distance from 675 to 775 as easily?

### Michael's Approach

Michael, who also easily finds the distance from 650 to 1000, then works on the more difficult problem, "How far is it from 480 to 740?"

At first he says, "It's 300 and something," as he looks at the hundreds place of both numbers. "Um ... maybe 340," he says, now focusing on the numbers in the tens place, "80 and 40 makes 120, so maybe you have to put a 20 with the 300 ... 320? 420?"

The teacher recognizes one of Michael's panics about getting the "right answer." Since he easily

*Continued on next page*

solved some of the easier distance problems, the teacher knows he has many of the skills he needs to solve this one. However, his focus on the numerals seems to be interfering with his good numerical reasoning strategies.

The teacher asks Michael to use the 1000 chart to check his answer. He puts one finger on 480 and one on 740.

**Now what are you going to do?**

**Michael:** Count.

**How are you going to count?**

**Michael:** By 1's.

Again, Michael is not calling on the skills he has used in other situations, so the teacher poses another problem.

**It's going to take too long to count by 1's, and it's too easy to make a mistake. What would you do if you wanted to find out how far it is from 60 to 90?**

**Michael** [*promptly*]: 70, 80, 90—that's 30.

**OK, so could you use 10's like that for this one?**

Michael then counts from 480 to 740 by tens, leaving one finger on 480 as he counts. He leaves out one number—"240, 260, 270"—so he ends up with 270 instead of 260. At this point, the teacher has learned enough about Michael's strategies and asks him to write down his solution, both for his portfolio and as a way of getting him to articulate and remember his approach.

Michael is able to operate in the 100's, knows how to count by 10's across multiples of 100, and knows where numbers in the 100's are in relation to each other. However, he is not yet using landmarks (such as 500 and 700, in this case) to help him figure out relationships between numbers. He seems not to be using any visual images to help him "see" the relationships between 480 and 740; rather he focuses on the numerals themselves, which in this case mislead him about the difference between the two numbers.

When using the 1000 chart, Michael still seems intimidated by larger numbers, works very hard to keep track of where he is, and seems worried about losing his place. The teacher decides he needs to work on more of these problems, using the 1000 chart, starting with pairs of numbers that include one multiple of 100. For example, "How far is it from 500 to 740?"

# Counting Around the Class

## Basic Activity

Students count around the class by a particular number. That is, if counting by 2's, the first student says "2," the next student says "4," the next "6," and so forth. Before the count starts, students try to predict on what number the count will end. During and after the count, students discuss relationships between the chosen factor and its multiples.

Counting Around the Class is designed to give students practice with counting by many different numbers and to foster numerical reasoning about the relationships among factors and their multiples. Students focus on:

■ becoming familiar with multiplication patterns

■ relating factors to their multiples

■ developing number sense about multiplication and division relationships

## Materials

■ Calculators (for variation)

## Procedure

**Step 1. Choose a number to count by.** For example, if the class has been working with quarters recently, you might want to count by 25's.

**Step 2. Ask students to predict the target number.** "If we count by 25's around the class, what number will we end up on?" Encourage students to talk about how they could figure this out without doing the actual counting.

**Step 3. Count around the class by your chosen number.** "25 ... 50 ... 75 ..." If some students seem uncertain about what comes next, you might put the numbers on the board as they count; seeing the visual patterns can help some students with the spoken one.

You might count around a second time by the same number, starting with a different person, so that students will hear the pattern more than once and have their turns at different points in the sequence.

**Step 4. Pause in the middle of the count to look back.** "We're up to 375 counting by 25's. How many students have we counted so far? How do you know?"

**Step 5. Extend the problem.** Ask questions like these: "Which of your predictions were reasonable? Which were possible? Which were impossible?" (A student might remark, for example, "You couldn't have 510 for 25's because 25 only lands on the 25's, the 50's, the 75's, and the 100's.")

"What if we had 32 students in this class instead of 28? Then where would we end up?"

"What if we used a different number? This time we counted by 25's and ended on 700; what if we counted by 50's? What number do you think we would end on? Why do you think it will be twice as big? How did you figure that out?"

## Variations

**Multiplication Practice** Use single-digit numbers to provide practice with multiplication (that is, count by 2's, 3's, 4's, 5's, 7's, and so forth). In counting by numbers other than 1, students usually first become comfortable with 2's, 5's, and 10's, which have very regular patterns. Soon they can begin to count by more difficult single-digit numbers: 3, 4, 6, and (later) 7, 8, and 9.

**Landmark Numbers** When students are learning about money or about our base ten system of numeration, they can count by 20's, 25's, 50's, 100's, and 1000's. Counting by multiples of 10 and 100 (30's, 40's, 600's) will support students' growing familiarity with the base 10 system.

*Continued on next page*

**Making Connections** When you choose harder numbers, pick those that are related in some way to numbers students are very familiar with. For example, once students are comfortable counting by 25's, have them count by 75's. Ask students how knowing the 25's will help them count by 75's. If students are fluent with 3's, try counting by 6's or by 30's. If students are fluent with 10's and 20's, start working on 15's. If they are comfortable counting by 15's, ask them to count by 150's or 1500's.

**Large Numbers** Introduce large numbers, such as 2000, 5000, 1500, or 10,000, so that students begin to work with combinations of these less familiar numbers.

**Using the Calculator** On some days you might have everyone use a calculator, or have a few students use the calculators to skip count while you are counting around the class. On most calculators, the equals (=) key provides a built-in constant function, allowing you to skip count easily. For example, if you want to skip count by 25's, you press your starting number (let's say 0), the operation you want to use (in this case, +), and the number you want to count by (in this case, 25). Then, press the equals key each time you want to add 25. So, if you press

| 0 | + | 25 | = | = | = | = |

you will see on your screen 25, 50, 75, 100.

## Special Notes
**Letting Students Prepare** When introducing an unfamiliar number to count with, students may need some preparation before they try to count around the class. Ask students to work in pairs to figure out, with whatever materials they want to use, what number the count will end on.

**Avoiding Competition** It is important to be sensitive to potential embarrassment or competition if some students have difficulty figuring out their number. One teacher allowed students to volunteer for the next number, rather than counting in a particular order. Other teachers have made the count a cooperative effort, establishing an atmosphere in which students readily helped each other, and anyone felt free to ask for help.

## Related Homework Options
**Counting Patterns** Students write out a counting pattern up to a target number (for example, by 25's up to 500). Then they write about what patterns they see in their counting. Calculators can be used for this.

**Mystery Number Problems** Provide an ending number and ask students to figure out what factor they would have to count by to reach it. For example: "I'm thinking of a mystery number. I figured out that if we counted around the class by my mystery number today, we would get to 2800. What is the mystery number?"

Or, you might provide students with the final number and the factor, and ask them to figure out the number of students in the class. "When a certain class counts by 25's, the last student says 550. How many students are in the class?" Calculators can be used.

# Calendar Math

## Basic Activity

Students try to find numerical expressions that are equal to the day's date. For example, if the date is March 19, students look for ways to combine numbers and operations to make 19. Constraints on what numbers or operations they can use push students in developing their arithmetic skills. Discoveries of principles for "expressions that work" become a part of the class's mathematics culture.

Calendar Math is a simple way of providing arithmetic practice and opportunities for students to share mathematical discoveries. Students focus on:

- developing a web of numerical knowledge about any number

- using operations flexibly

- recognizing relationships among operations (for example, that adding and then subtracting the same number has a net effect of 0)

- learning about and using key mathematical ideas, such as the effect of using an operation with 0 or 1

- deriving new numerical expressions by modifying a particular expression systematically (for example, if $2 + 9 = 11$, then so does $3 + 8$, $4 + 7$, $5 + 6$, and so forth)

## Materials

- Calculators (for variation)

## Procedure

**Step 1. Pose the problem.** For example, "Today's date is September 12. Who can think of a way we could combine numbers to make 12?"

**Step 2. List student responses.** Their ideas might include expressions like these:

| | | |
|---|---|---|
| $6 + 6$ | $4 \times 3$ | $12 + 0$ |
| $1/2 \times 24$ | $3 + (-3) + 6 + 6$ | |

**Step 3. Choose a "favorite expression" for the day.** Students choose their favorite from the listed expressions, perhaps the most unusual, or one that uses a new idea. Use the class favorite to write the date on the board: September $(0 \times 12) + 12$.

## Variations

**Introducing Constraints** Introduce constraints based on your class's ease with particular operations and numbers. For example, if students are very comfortable with addition, eliminate addition as a possibility: "Today you can use any operation you want to use, *except* addition." You could also require that a certain operation or kind of number is used. Possible constraints include these:

> You can't use any number that's a multiple of 2.
>
> You can't use addition or subtraction.
>
> You must use more than one operation.
>
> You must use one number that's bigger than 100 (1000, 5000).
>
> You must start with 100.
>
> You must use at least three numbers.
>
> You can't use 0.
>
> You must use at least one number that is smaller than 1.
>
> You must use one negative number.
>
> You can only use 1's, 2's, 3's, and 4's.

**Looking for Patterns** Encourage students to find expressions that they can alter systematically to find more expressions. Here is a pattern that a student came up with for 20:

> $2 \times 10$, $2 \times 9 + 2$, $2 \times 8 + 4$, $2 \times 7 + 6$ ...

**Developing Class "Rules"** Our experience is that, through this activity, new ideas about numbers become part of the culture of the classroom. For example, in one classroom, one student learned about square numbers and the notation for them. Because she used these numbers in

*Continued on next page*

Calendar Math, other students became familiar with them and were soon using numbers such as $4^2$.

Other kinds of relationships are often discovered by one child and then become common knowledge. For example, one day when a class was finding expressions equivalent to 14, one student suggested "14 times 0." This remark prompted a discussion of what happens when you multiply a number by 0, and students eventually concluded that the result of multiplying any number by zero is 0. The teacher wrote this on a list of "rules" discovered over the course of the year. Following are some other rules students have discovered during this activity:

A number divided by 1 is the number.

A number multiplied by 1 is the number.

Any number multiplied by 0 is 0.

Any number divided by itself is 1.

Subtract or add 0 to any number and you still have the same number.

Adding lots of 0's doesn't change anything.

You can make any number by adding enough 1's to count up to that number.

Adding a number and then subtracting the same number is like adding 0.

**Using the Calculator**   For a quiet ten minutes, have students work individually or in pairs on coming up with ways to make the date, using their calculators. Make sure that they record their work on a piece of paper. They can choose their favorite solution and write it on the board. This is a good way for students to explore new keys on the calculator.

## Related Homework Options

**Planning Ahead**   Suggest that students think at home about how they might make the next day's date. Tell students what constraints they are under, and ask them to figure out five different ways to make the date. They can share their favorite in class.

The following activities will help ensure that this unit is comprehensible to students who are acquiring English as a second language. The suggested approach is based on *The Natural Approach: Language Acquisition in the Classroom* by Stephen D. Krashen and Tracy D. Terrell (Alemany Press, 1983). The intent is for second-language learners to acquire new vocabulary in an active, meaningful context.

Note that *acquiring* a word is different from *learning* a word. Depending on their level of proficiency, students may be able to comprehend a word upon hearing it during an investigation, without being able to say it. Other students may be able to use the word orally, but not read or write it. The goal is to help students naturally acquire targeted vocabulary at their present level of proficiency.

We suggest using these activities just before the related investigations. The activities can also be led by English-proficient students.

## Investigations 1–3

*the numbers 1 to 100*

1. Create action commands that ask students to nonverbally identify numbers written on the board.

   **Put your finger on the number 6.**

   **Cover the number 13.**

2. Use classroom items (paper clips, rubber bands) to help students count to 100.

*divide, equal, unequal, group*

1. Divide 10 pencils in two equal groups. Identify the piles as equal by showing how each group has an *equal* number of pencils.

2. Divide the pencils into two unequal piles. Shake your head as you explain that these groups have an *unequal* number of pencils.

3. Repeat the same procedure, using other classroom items (books, paper clips).

4. Challenge students to demonstrate comprehension of these words by following action commands.

   **Divide these pencils into two equal groups.**

   **Divide these paper clips into three equal groups.**

   **Divide these books into four unequal groups.**

*money: coins, cents, nickel, dime, quarter, dollar*

1. Use a dollar, quarter, dime, and nickel along with action commands to help familiarize students with these words.

   **Put a quarter in your hand.**

   **Give me a dime.**

   **Put the nickel under the dollar.**

   **Put all the coins in a pile.**

   **Fold the dollar in half.**

2. Create action commands that require students to identify coins by their value.

   **Take a coin that is worth 5 cents.**

   **Give me a coin that is worth 10 cents.**

   **Show me which coin is worth the most money.**

*pattern*

1. Start a clapping pattern (hit knees, hit knees, clap; hit knees, hit knees, clap). Ask students to join in. Then change the pattern.

   **Try this pattern with me.**

   **Who wants to show us a new pattern?**

2. Give students two colors of interlocking cubes. Show them a pattern of cubes in a line, such as blue, red, red, blue, red, red. Ask students to make their own patterns.

3. Draw or write patterns on the board, identify them, and ask students to continue them.

   **This is a pattern of shapes; what comes next?**

   □ △ ○ □ △ ○ □ △ ○

   **This is a pattern in numbers; what comes next?** [*Write the numbers in order, then circle and say aloud each even number, pausing at 8.*]

   1 ② 3 ④ 5 ⑥ 7 ⑧ 9  10  11  12 ...

   **This is another kind of number pattern; what comes next?** [*Write the numbers in a column.*]

   3
   13
   23
   33
   43

# Blackline Masters

_____ , 19 ___

# Dear Family,

In mathematics, our class is starting a new unit called *Landmarks in the Hundreds*. This unit will help your child learn about important numbers like 100 and 1000, that we use to find our way around the number system.

First your child will be working with 100—an especially important landmark. Then we'll work with multiples of 100, 200, 300, 400, and so on up to 1000. We'll be doing "skip counting"—that's counting by 2's, or 5's, or 10's, or any other number. The children will work in class, and sometimes at home, doing problems like this one:

> If you count by 2's (2, 4, 6, 8, 10, 12, 14 ...) you eventually land on 100. What other numbers can you count by that will land you exactly on 100? What numbers don't work? How many 2's did you count to get to 100?

Children have very interesting ways to figure out these problems. You can help by asking your child to tell you how he or she got an answer. There are many ways of doing these problems—and no single "right" way. What's important for your child to know is how his or her own way works. This is all part of developing good common sense about numbers.

We'll also be doing estimation. I may ask your child to find groups of things around the house that number about 100, about 200, about 300, and so forth. You can help in your child's search. Talk about the number of things you both are finding, for example:

> What about the nuts in this jar? Do you think that would be close to 100? What about the squares in the ceiling?

Finally, any time that you yourself need to estimate or deal with large numbers, please involve your child. Whether you're buying food, or deciding how many tiles to buy to patch the floor, your child probably has some good ideas about how to go about it.

Sincerely,

# Ways to Make 20

Take 20 cubes. Put them in equal groups.
How many different ways can you make 20?
Record your results here.

| **Numbers I tried that will make 20** |
|---|
| Number of cubes in group: _____<br>Picture of 20:<br><br><br><br>Count: |
| Number of cubes in group: _____<br>Picture of 20:<br><br><br><br>Count: |
| Number of cubes in group: _____<br>Picture of 20:<br><br><br><br>Count: |
| **Numbers I tried that didn't work** |
| <br><br> |

# Ways to Make _____

Choose a number to write in the blank above.
Find equal groupings that make your number.
Record your results in the chart.

**Note:** In class, we use cubes to make groupings. At home you might use beans, popcorn, pennies, paper clips, pebbles, coins, or dots drawn on paper.

| **The number I am making is ____** |
|---|
| Number of cubes in each group: _____<br>Picture of how you made your number with these groups<br><br><br>Skip count by the number of cubes in each group. |
| Number of cubes in each group: _____<br>Picture of how you made your number with these groups<br><br><br>Skip count by the number of cubes in each group. |
| Number of cubes in each group: _____<br>Picture of how you made your number with these groups<br><br><br>Skip count by the number of cubes in each group. |
| **Numbers I tried that didn't work** |
|  |

# Factors of 24

| 1 | 2 | 3 | 4 | 5 | 6 | 7 | 8 | 9 | 10 |
|---|---|---|---|---|---|---|---|---|----|
| 11 | 12 | 13 | 14 | 15 | 16 | 17 | 18 | 19 | 20 |
| 21 | 22 | 23 | 24 | | | | | | |

_____ is a factor of 24.

How many do you need to make 24? _____

_____ × _____ = 24

| 1 | 2 | 3 | 4 | 5 | 6 | 7 | 8 | 9 | 10 |
|---|---|---|---|---|---|---|---|---|----|
| 11 | 12 | 13 | 14 | 15 | 16 | 17 | 18 | 19 | 20 |
| 21 | 22 | 23 | 24 | | | | | | |

_____ is a factor of 24.

How many do you need to make 24? _____

_____ × _____ = 24

| 1 | 2 | 3 | 4 | 5 | 6 | 7 | 8 | 9 | 10 |
|---|---|---|---|---|---|---|---|---|----|
| 11 | 12 | 13 | 14 | 15 | 16 | 17 | 18 | 19 | 20 |
| 21 | 22 | 23 | 24 | | | | | | |

_____ is a factor of 24.

How many do you need to make 24? _____

_____ × _____ = 24

| 1 | 2 | 3 | 4 | 5 | 6 | 7 | 8 | 9 | 10 |
|---|---|---|---|---|---|---|---|---|----|
| 11 | 12 | 13 | 14 | 15 | 16 | 17 | 18 | 19 | 20 |
| 21 | 22 | 23 | 24 | | | | | | |

_____ is a factor of 24.

How many do you need to make 24? _____

_____ × _____ = 24

# Factors of 36

| 1 | 2 | 3 | 4 | 5 | 6 | 7 | 8 | 9 | 10 |
|---|---|---|---|---|---|---|---|---|----|
| 11 | 12 | 13 | 14 | 15 | 16 | 17 | 18 | 19 | 20 |
| 21 | 22 | 23 | 24 | 25 | 26 | 27 | 28 | 29 | 30 |
| 31 | 32 | 33 | 34 | 35 | 36 | | | | |

_____ is a factor of 36.

How many do you need to make 36? _____

_____ × _____ = 36

| 1 | 2 | 3 | 4 | 5 | 6 | 7 | 8 | 9 | 10 |
|---|---|---|---|---|---|---|---|---|----|
| 11 | 12 | 13 | 14 | 15 | 16 | 17 | 18 | 19 | 20 |
| 21 | 22 | 23 | 24 | 25 | 26 | 27 | 28 | 29 | 30 |
| 31 | 32 | 33 | 34 | 35 | 36 | | | | |

_____ is a factor of 36.

How many do you need to make 36? _____

_____ × _____ = 36

| 1 | 2 | 3 | 4 | 5 | 6 | 7 | 8 | 9 | 10 |
|---|---|---|---|---|---|---|---|---|----|
| 11 | 12 | 13 | 14 | 15 | 16 | 17 | 18 | 19 | 20 |
| 21 | 22 | 23 | 24 | 25 | 26 | 27 | 28 | 29 | 30 |
| 31 | 32 | 33 | 34 | 35 | 36 | | | | |

_____ is a factor of 36.

How many do you need to make 36? _____

_____ × _____ = 36

# Factors of 48

| 1 | 2 | 3 | 4 | 5 | 6 | 7 | 8 | 9 | 10 |
|---|---|---|---|---|---|---|---|---|----|
| 11 | 12 | 13 | 14 | 15 | 16 | 17 | 18 | 19 | 20 |
| 21 | 22 | 23 | 24 | 25 | 26 | 27 | 28 | 29 | 30 |
| 31 | 32 | 33 | 34 | 35 | 36 | 37 | 38 | 39 | 40 |
| 41 | 42 | 43 | 44 | 45 | 46 | 47 | 48 | | |

_____ is a factor of 48.

How many do you need to make 48? _____

_____ × _____ = 48

| 1 | 2 | 3 | 4 | 5 | 6 | 7 | 8 | 9 | 10 |
|---|---|---|---|---|---|---|---|---|----|
| 11 | 12 | 13 | 14 | 15 | 16 | 17 | 18 | 19 | 20 |
| 21 | 22 | 23 | 24 | 25 | 26 | 27 | 28 | 29 | 30 |
| 31 | 32 | 33 | 34 | 35 | 36 | 37 | 38 | 39 | 40 |
| 41 | 42 | 43 | 44 | 45 | 46 | 47 | 48 | | |

_____ is a factor of 48.

How many do you need to make 48? _____

_____ × _____ = 48

| 1 | 2 | 3 | 4 | 5 | 6 | 7 | 8 | 9 | 10 |
|---|---|---|---|---|---|---|---|---|----|
| 11 | 12 | 13 | 14 | 15 | 16 | 17 | 18 | 19 | 20 |
| 21 | 22 | 23 | 24 | 25 | 26 | 27 | 28 | 29 | 30 |
| 31 | 32 | 33 | 34 | 35 | 36 | 37 | 38 | 39 | 40 |
| 41 | 42 | 43 | 44 | 45 | 46 | 47 | 48 | | |

_____ is a factor of 48.

How many do you need to make 48? _____

_____ × _____ = 48

# Miniature 100 Charts

| 1 | 2 | 3 | 4 | 5 | 6 | 7 | 8 | 9 | 10 |
|---|---|---|---|---|---|---|---|---|---|
| 11 | 12 | 13 | 14 | 15 | 16 | 17 | 18 | 19 | 20 |
| 21 | 22 | 23 | 24 | 25 | 26 | 27 | 28 | 29 | 30 |
| 31 | 32 | 33 | 34 | 35 | 36 | 37 | 38 | 39 | 40 |
| 41 | 42 | 43 | 44 | 45 | 46 | 47 | 48 | 49 | 50 |
| 51 | 52 | 53 | 54 | 55 | 56 | 57 | 58 | 59 | 60 |
| 61 | 62 | 63 | 64 | 65 | 66 | 67 | 68 | 69 | 70 |
| 71 | 72 | 73 | 74 | 75 | 76 | 77 | 78 | 79 | 80 |
| 81 | 82 | 83 | 84 | 85 | 86 | 87 | 88 | 89 | 90 |
| 91 | 92 | 93 | 94 | 95 | 96 | 97 | 98 | 99 | 100 |

I counted by _____.

I made _____ jumps to reach 100.

| 1 | 2 | 3 | 4 | 5 | 6 | 7 | 8 | 9 | 10 |
|---|---|---|---|---|---|---|---|---|---|
| 11 | 12 | 13 | 14 | 15 | 16 | 17 | 18 | 19 | 20 |
| 21 | 22 | 23 | 24 | 25 | 26 | 27 | 28 | 29 | 30 |
| 31 | 32 | 33 | 34 | 35 | 36 | 37 | 38 | 39 | 40 |
| 41 | 42 | 43 | 44 | 45 | 46 | 47 | 48 | 49 | 50 |
| 51 | 52 | 53 | 54 | 55 | 56 | 57 | 58 | 59 | 60 |
| 61 | 62 | 63 | 64 | 65 | 66 | 67 | 68 | 69 | 70 |
| 71 | 72 | 73 | 74 | 75 | 76 | 77 | 78 | 79 | 80 |
| 81 | 82 | 83 | 84 | 85 | 86 | 87 | 88 | 89 | 90 |
| 91 | 92 | 93 | 94 | 95 | 96 | 97 | 98 | 99 | 100 |

I counted by _____.

I made _____ jumps to reach 100.

| 1 | 2 | 3 | 4 | 5 | 6 | 7 | 8 | 9 | 10 |
|---|---|---|---|---|---|---|---|---|---|
| 11 | 12 | 13 | 14 | 15 | 16 | 17 | 18 | 19 | 20 |
| 21 | 22 | 23 | 24 | 25 | 26 | 27 | 28 | 29 | 30 |
| 31 | 32 | 33 | 34 | 35 | 36 | 37 | 38 | 39 | 40 |
| 41 | 42 | 43 | 44 | 45 | 46 | 47 | 48 | 49 | 50 |
| 51 | 52 | 53 | 54 | 55 | 56 | 57 | 58 | 59 | 60 |
| 61 | 62 | 63 | 64 | 65 | 66 | 67 | 68 | 69 | 70 |
| 71 | 72 | 73 | 74 | 75 | 76 | 77 | 78 | 79 | 80 |
| 81 | 82 | 83 | 84 | 85 | 86 | 87 | 88 | 89 | 90 |
| 91 | 92 | 93 | 94 | 95 | 96 | 97 | 98 | 99 | 100 |

I counted by _____.

I made _____ jumps to reach 100.

# Factors of 100

Use this recording sheet with the Miniature 100 Charts.

| Numbers we tried | Did you land on 100 exactly? | If it worked: How many in 100? |
|---|---|---|
| _____ | Yes  No | _____ |
| _____ | Yes  No | _____ |
| _____ | Yes  No | _____ |
| _____ | Yes  No | _____ |
| _____ | Yes  No | _____ |
| _____ | Yes  No | _____ |
| _____ | Yes  No | _____ |
| _____ | Yes  No | _____ |
| _____ | Yes  No | _____ |
| _____ | Yes  No | _____ |
| _____ | Yes  No | _____ |
| _____ | Yes  No | _____ |
| _____ | Yes  No | _____ |
| _____ | Yes  No | _____ |
| _____ | Yes  No | _____ |
| _____ | Yes  No | _____ |
| _____ | Yes  No | _____ |
| _____ | Yes  No | _____ |
| _____ | Yes  No | _____ |
| _____ | Yes  No | _____ |

# A Picture of 100

Make a picture of 100 of the same thing.

You might draw 100 things.
Or you might glue 100 of something small
onto a sheet of paper.

Arrange the 100 things in equal groups.
Make it easy to see that you have 100 things
on your paper.

Remember the picture of 100 you made in class?
Use groups of a **different** size.

Your family can help you make this picture.

# Ways to Split Up a Dollar

| Number of People Sharing a Dollar | How Much Each Person Gets |
|---|---|
|  |  |
|  |  |
|  |  |
|  |  |
|  |  |
|  |  |
|  |  |
|  |  |
|  |  |
|  |  |
|  |  |

Cannot split a dollar evenly:

# Share a Dollar

Solve the following problem about splitting up a dollar.

Ten third grade students were playing out at recess, and they found $1.00. No one claimed the $1.00, and their teacher told them that if they could figure out a fair way to split up the money, they could keep it. How did they split it up? How did you get your answer?

Now write your own problem about splitting up a dollar.

On the back of this sheet, solve your own problem. Be prepared to share your problem with some of the students in your class.

# Exploring Multiples of 100

Choose one of these factors
of 100. Circle it.

4    5    10    20    25

Then fill in the chart.

Show how many of your
number makes 100, 200,
and 300.

You can go further if
you want to.

|  | How many? |
|---|---|
| 100 | _____ |
| 200 | _____ |
| 300 | _____ |
| _____ | _____ |
| _____ | _____ |
| _____ | _____ |
| _____ | _____ |
| _____ | _____ |
| _____ | _____ |

I know there are _____ in 300 because_____

_____

_____

_____

_____

This is what we notice about the patterns in our list:

_____

_____

_____

# Calculator Skip Counting

Choose a number to count by. Pick one you think will land exactly on 300.

Skip count by this number on your calculator.

Does it work? If so, write how many of your number it takes to get to 300.

| Numbers we tried | Did you land on 300 exactly? | | If it worked: How many in 300? |
|---|---|---|---|
| _____ | Yes | No | _____ |
| _____ | Yes | No | _____ |
| _____ | Yes | No | _____ |
| _____ | Yes | No | _____ |
| _____ | Yes | No | _____ |
| _____ | Yes | No | _____ |
| _____ | Yes | No | _____ |
| _____ | Yes | No | _____ |
| _____ | Yes | No | _____ |
| _____ | Yes | No | _____ |
| _____ | Yes | No | _____ |
| _____ | Yes | No | _____ |
| _____ | Yes | No | _____ |
| _____ | Yes | No | _____ |
| _____ | Yes | No | _____ |
| _____ | Yes | No | _____ |
| _____ | Yes | No | _____ |
| _____ | Yes | No | _____ |
| _____ | Yes | No | _____ |

# How Many 4's Are in 600?

There are _____ 4's in 600. Here is how
I figured this out.

_____

_____

_____

_____

_____

Draw a picture to show how you figured this out.

# Multiplying Things in Class

Find something in the classroom that comes in groups.
Each group should have the same amount.
Figure out how many there are altogether in the classroom.

Try to find things that total in the hundreds.
Are there nearly 200? nearly 300? nearly 400?

For example, how many fingers are there in the room?
Count by 10's to find the total.

Here are some other things you might look for:

- boxes of paper clips
- packages or pads of paper
- packs of pencils
- chair and desk legs
- rows of bricks or tiles

# Multiplying Things at Home

Find something at home that comes in groups.
Each group should have the same amount.
Figure out how many there are altogether.

Try to find things that total in the hundreds.
Are there nearly 200? nearly 300? nearly 400?

Count by 10's to find the total.

Here are some things you might look for:

- windowpanes in your windows
- fingers in your family
- spokes on bicycle wheels belonging to your family and friends
- chair legs
- rows of bricks or tiles

Write about some of the things you found, and about how you figured out how many of them there were.

_____

_____

_____

_____

# Making Up Landmark Problems

Make up three problems using landmarks. Follow these rules:

1. Use important landmark numbers such as 20, 25, 75, 100, and so forth.

2. Each problem must be a real situation or something that could happen.

3. Solve every problem you write. Use a separate piece of paper for each solution.

After you have solved the problems, ask someone else in your family to solve them too. See if they know how to use landmark numbers.

**MONEY PROBLEMS**

Write how you solved each problem on a separate piece of paper.

| | |
|---|---|
| How many quarters are there in $3.75? | How many quarters are there in $4.25? |
| How much money do I have if I have 17 quarters? | How much money is 23 quarters? |
| How many dimes are there in $4.30? | How much money is 13 quarters? |
| How much money is 27 dimes? | How many nickels are there in $3.75? |

You have $3.75. Pencils cost a quarter each.

How many pencils can you buy?

Ten people got on the bus. Each paid 75 cents.

How much did the bus driver collect from them?

The early show at the movies costs $2.50.

How much will four friends pay to go to the early show?

Someone had $6.00 in two pockets. There was $3.75 in one pocket.

How much was in the other pocket?

Suppose each person in class today had a quarter in his or her pocket.

How much money would we have?

I want to bring my friends with me to the swimming pool. The pool charges 75 cents for each person. I saved up $9.75.

How many friends can I invite?

There are 380 students in a school. There are about 20 students in each class.

How many classes are there?

---

I counted the wheels in the school parking lot. There were 84 wheels.

How many cars were in the parking lot?

---

I'm baking cookies. I want to make about 250 cookies. I can bake 25 cookies in a batch.

How many batches should I make?

---

Some students had a cookie sale. They sold chocolate chip cookies for 5 cents each. They earned $4.60.

How many cookies did they sell?

---

There are 390 students in a school. There are about 30 students in each class.

How many classes are there?

---

Two friends had a lemonade stand. They charged 15 cents for each glass. One day, they sold enough to get $1.80.

How many glasses did they sell?

# More Than 300

Look for quantities of things at home that match
one or more of the numbers on your 1000 chart.
The number(s) should be greater than 300 and
less than 1000.

| What I found | How many there were |
|---|---|
|  |  |
|  |  |
|  |  |
|  |  |
|  |  |

| 1 | 2 | 3 | 4 | 5 | 6 | 7 | 8 | 9 | 10 |
|---|---|---|---|---|---|---|---|---|---|
| 11 | 12 | 13 | 14 | 15 | 16 | 17 | 18 | 19 | 20 |
| 21 | 22 | 23 | 24 | 25 | 26 | 27 | 28 | 29 | 30 |
| 31 | 32 | 33 | 34 | 35 | 36 | 37 | 38 | 39 | 40 |
| 41 | 42 | 43 | 44 | 45 | 46 | 47 | 48 | 49 | 50 |
| 51 | 52 | 53 | 54 | 55 | 56 | 57 | 58 | 59 | 60 |
| 61 | 62 | 63 | 64 | 65 | 66 | 67 | 68 | 69 | 70 |
| 71 | 72 | 73 | 74 | 75 | 76 | 77 | 78 | 79 | 80 |
| 81 | 82 | 83 | 84 | 85 | 86 | 87 | 88 | 89 | 90 |
| 91 | 92 | 93 | 94 | 95 | 96 | 97 | 98 | 99 | 100 |

| 1 | 2 | 3 | 4 | 5 | 6 | 7 | 8 | 9 | 10 |
|---|---|---|---|---|---|---|---|---|---|
| 11 | 12 | 13 | 14 | 15 | 16 | 17 | 18 | 19 | 20 |
| 21 | 22 | 23 | 24 | 25 | 26 | 27 | 28 | 29 | 30 |
| 31 | 32 | 33 | 34 | 35 | 36 | 37 | 38 | 39 | 40 |
| 41 | 42 | 43 | 44 | 45 | 46 | 47 | 48 | 49 | 50 |
| 51 | 52 | 53 | 54 | 55 | 56 | 57 | 58 | 59 | 60 |
| 61 | 62 | 63 | 64 | 65 | 66 | 67 | 68 | 69 | 70 |
| 71 | 72 | 73 | 74 | 75 | 76 | 77 | 78 | 79 | 80 |
| 81 | 82 | 83 | 84 | 85 | 86 | 87 | 88 | 89 | 90 |
| 91 | 92 | 93 | 94 | 95 | 96 | 97 | 98 | 99 | 100 |
| 101 | 102 | 103 | 104 | 105 | 106 | 107 | 108 | 109 | 110 |
| 111 | 112 | 113 | 114 | 115 | 116 | 117 | 118 | 119 | 120 |
| 121 | 122 | 123 | 124 | 125 | 126 | 127 | 128 | 129 | 130 |
| 131 | 132 | 133 | 134 | 135 | 136 | 137 | 138 | 139 | 140 |
| 141 | 142 | 143 | 144 | 145 | 146 | 147 | 148 | 149 | 150 |
| 151 | 152 | 153 | 154 | 155 | 156 | 157 | 158 | 159 | 160 |
| 161 | 162 | 163 | 164 | 165 | 166 | 167 | 168 | 169 | 170 |
| 171 | 172 | 173 | 174 | 175 | 176 | 177 | 178 | 179 | 180 |
| 181 | 182 | 183 | 184 | 185 | 186 | 187 | 188 | 189 | 190 |
| 191 | 192 | 193 | 194 | 195 | 196 | 197 | 198 | 199 | 200 |
| 201 | 202 | 203 | 204 | 205 | 206 | 207 | 208 | 209 | 210 |
| 211 | 212 | 213 | 214 | 215 | 216 | 217 | 218 | 219 | 220 |
| 221 | 222 | 223 | 224 | 225 | 226 | 227 | 228 | 229 | 230 |
| 231 | 232 | 233 | 234 | 235 | 236 | 237 | 238 | 239 | 240 |
| 241 | 242 | 243 | 244 | 245 | 246 | 247 | 248 | 249 | 250 |
| 251 | 252 | 253 | 254 | 255 | 256 | 257 | 258 | 259 | 260 |
| 261 | 262 | 263 | 264 | 265 | 266 | 267 | 268 | 269 | 270 |
| 271 | 272 | 273 | 274 | 275 | 276 | 277 | 278 | 279 | 280 |
| 281 | 282 | 283 | 284 | 285 | 286 | 287 | 288 | 289 | 290 |
| 291 | 292 | 293 | 294 | 295 | 296 | 297 | 298 | 299 | 300 |

# Practice Pages

This optional section provides homework ideas for teachers who want or need to give more homework than is assigned to accompany the activities in this unit. The problems included here provide additional practice in learning about number relationships and in solving computation and number problems. For number units, you may want to use some of these if your students need more work in these areas or if you want to assign daily homework. For other units, you can use these problems so that students can continue to work on developing number and computation sense while they are focusing on other mathematical content in class. We recommend that you introduce activities in class before assigning related problems for homework.

**The Arranging Chairs Puzzle**    This activity is introduced in the unit *Things That Come in Groups*. If your students are familiar with the activity, you can simply send home the directions so that students can play at home. If your students have not done this activity before, introduce it in class and have students do it once or twice before sending it home. Early in the year, ask students to work with numbers such as 15, 18, and 24. Later in the year, as they become ready to work with larger numbers, they can try numbers such as 32, 42, or 50. You might have students do this activity two times for homework in this unit.

**Money Problems**    This type of problem is introduced in the unit *Mathematical Thinking at Grade 3*. Here you are provided two of these problems for student homework. You can make up other problems in this format, using numbers that are appropriate for your students. Students record their strategies for solving the problems, using numbers, words, or pictures.

**How Many Legs?**    This type of problem is introduced in the unit *Things That Come in Groups*. Provided here are two such problems for student homework. You can make up other problems in this format, using numbers that are appropriate for your students. Students record their strategies for solving the problems, using numbers, words, or pictures.

# The Arranging Chairs Puzzle

## What You Will Need

30 small objects to use as chairs (for example, cubes, blocks, tiles, chips, pennies, buttons)

## What to Do

1. Choose a number between 4 and 30.

2. Figure out all the ways you can arrange that many chairs. Each row must have the same number of chairs. Your arrangements will make rectangles of different sizes.

3. Write down the dimensions of each rectangle you make.

4. Choose another number and start again. Be sure to make a new list of dimensions for each new number.

### Example
All the ways to arrange
12 chairs

### Dimensions
1 by 12
12 by 1
2 by 6
6 by 2
3 by 4
4 by 3

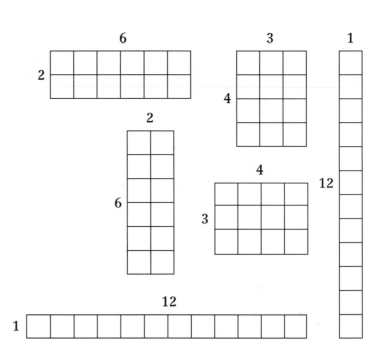

# Practice Page A

I have a total of 75¢ in my 2 pockets. I have 8 nickels in 1 pocket. What coins could I have in my other pocket?

Show how you solved this problem. You can use numbers, words, or
pictures.

## Practice Page B

Jennifer and I sold cookies. We made $1.60. We want to split the money evenly. How can we do this?

Show how you solved this problem. You can use numbers, words, or pictures.

# Practice Page C

Show how you solved each problem. You can use numbers, words, or pictures.

Eight friends are sitting together in a circle.

How many legs are there?

How many legs and arms are there?

How many legs and arms are there in a circle of 24 friends?

## Practice Page D

Show how you solved each problem. You can use numbers, words, or pictures.

People usually have 5 fingers on each hand.

How many fingers do 3 people have?




How many fingers do 7 people have?




How many fingers do 17 people have?